Journal of Applied Logics - IfCoLog Journal of Logics and their Applications

Volume 5, Number 4

June 2018

Disclaimer

ISBN 978-1-84890-272-5
ISSN (E) 2055-3714
ISSN (P) 2055-3706

College Publications
Scientific Director: Dov Gabbay
Managing Director: Jane Spurr

http://www.collegepublications.co.uk

Printed by Lightning Source, Milton Keynes, UK

EDITORIAL BOARD

Modal and Temporal Logic
Carlos Areces
Melvin Fitting
Victor Marek
Mark Reynolds.
Frank Wolter
Michael Zakharyaschev

Automated Inference Systems and Model Checking
Ed Clarke
Ulrich Furbach
Hans Juergen Ohlbach
Volker Sorge
Andrei Voronkov
Toby Walsh

Formal Methods: Specification and Verification
Howard Barringer
David Basin
Dines Bjorner
Kokichi Futatsugi
Yuri Gurevich

Logic and Software Engineering
Manfred Broy
John Fitzgerald
Kung-Kiu Lau
Tom Maibaum
German Puebla

Logic and Constraint Logic Programming
Manuel Hermenegildo
Antonis Kakas
Francesca Rossi
Gert Smolka

Logic and Databases
Jan Chomicki
Enrico Franconi
Georg Gottlob
Leonid Libkin
Franz Wotawa

Logic and Physics (space time. relativity and quantum theory)
Hajnal Andreka
Kurt Engesser
Daniel Lehmann
lstvan Nemeti
Victor Pambuccian

Logic for Knowledge Representation and the Semantic Web
Franz Baader
Anthony Cohn
Pat Hayes
Ian Horrocks
Maurizio Lenzerini
Bernhard Nebel

Tactical Theorem Proving and Proof Planning
Alan Bundy
Amy Felty
Jacques Fleuriot
Dieter Hutter
Manfred Kerber
Christoph Kreitz

Logic and Algebraic Programming
Jan Bergstra
John Tucker

Logic in Mechanical and Electrical Engineering
Rudolf Kruse
Ebrahaim Mamdani

Logic and Law
Jose Carmo
Lars Lindahl
Marek Sergot

Applied Non-classical Logic
Luis Farinas del Cerro
Nicola Olivetti

Mathematical Logic
Wilfrid Hodges
Janos Makowsky

Cognitive Robotics: Actions and Causation
Gerhard Lakemeyer
Michael Thielscher

Type Theory for Theorem Proving Systems
Peter Andrews
Chris Benzmüller
Chad Brown
Dale Miller
Carsten Schlirmann

Logic Applied in Mathematics (including e-Learning Tools for Mathematics and Logic)
Bruno Buchberger
Fairouz Kamareddine
Michael Kohlhase

Logic and Computational Models of Scientific Reasoning
Lorenzo Magnani
Luis Moniz Pereira
Paul Thagard

Logic and Multi-Agent Systems
Michael Fisher
Nick Jennings
Mike Wooldridge

Logic and Neural Networks
Artur d'Avila Garcez
Steffen Holldobler
John G. Taylor

Logic and Planning
Susanne Biundo
Patrick Doherty
Henry Kautz
Paolo Traverso

Algebraic Methods in Logic
Miklos Ferenczi
Robin Hirsch
Idiko Sain

Non-monotonic Logics and Logics of Change
Jurgen Dix
Vladimir Lifschitz
Donald Nute
David Pearce

Logic and Learning
Luc de Raedt
John Lloyd
Steven Muggleton

Logic and Natural Language Processing
Wojciech Buszkowski
Hans Kamp
Marcus Kracht
Johanna Moore
Michael Moortgat
Manfred Pinkal
Hans Uszkoreit

Fuzzy Logic Uncertainty and Probability
Didier Dubois
Petr Hajek
Jeff Paris
Henri Prade
George Metcalfe
Jon Williamson

Scope and Submissions

This journal considers submission in all areas of pure and applied logic, including:

pure logical systems
proof theory
constructive logic
categorical logic
modal and temporal logic
model theory
recursion theory
type theory
nominal theory
nonclassical logics
nonmonotonic logic
numerical and uncertainty reasoning
logic and AI
foundations of logic programming
belief revision
systems of knowledge and belief
logics and semantics of programming
specification and verification
agent theory
databases
dynamic logic
quantum logic
algebraic logic
logic and cognition
probabilistic logic
logic and networks
neuro-logical systems
complexity
argumentation theory
logic and computation
logic and language
logic engineering
knowledge-based systems
automated reasoning
knowledge representation
logic in hardware and VLSI
natural language
concurrent computation
planning

This journal will also consider papers on the application of logic in other subject areas: philosophy, cognitive science, physics etc. provided they have some formal content.

Submissions should be sent to Jane Spurr (jane.spurr@kcl.ac.uk) as a pdf file, preferably compiled in LaTeX using the IFCoLog class file.

CONTENTS

ARTICLES

Adams' p-validity in the research on human reasoning

Gernot D. Kleiter
Fachbereich Psychologie, Universität Salzburg, A-5020 Salzburg, Hellbrunnerstr. 34
gernot.kleiter@gmail.com, gernot.kleiter@sbg.ac.at

Abstract

This contribution throws a critical light on the application of Adams' probabilistic validity (p-validity) in the research on human reasoning. While Adams introduced "p-validity" in probability logic as a surrogate for "validity" in classical logic, it has recently been used in psychology as a "new standard" to evaluate probabilistic inferences. A major misunderstanding concerns the fact that in the work of Adams the probabilities of the premises are interval probabilities, while in the psychological experiments the participants assess point probabilities. The contribution argues that the coherence approach to probability, that is, the current continuation and extension of the work of de Finetti, is the more fruitful approach to evaluate and model human probabilistic inferences.

Keywords: p-validity, human reasoning, probability logic, coherence

1 Introduction

For more than a millennium philosophers compared human reasoning with logical principles. In the last century psychologists developed new theories and methods but continued to compare human inferences with the standards of classical logic. In the last two decades, however, a substantial number of psychologists working on human reasoning switched the perspective from classical logic to probability so that the old standards were not applicable any more. This included one of the most important standards of classical logic, the *validity* of inference rules:

> *If* $\boldsymbol{\phi} = \{\phi_1, \ldots, \phi_n\}$ *denotes a set of premises and* ψ *a conclusion, then an inference rule is valid,* $\boldsymbol{\phi} \models \psi$*, if and only if it is impossible for all premises in* $\boldsymbol{\phi}$ *to be true and the conclusion* ψ *to be false.*

I thank David Over for his help and patience in discussing many points of the paper. Thanks are due to the anonymous reviewers who helped to remove errors in the original manuscript and who stimulated a series of improvements.

Looking for a similar standard that applies to the probabilistic approach psychologists hit on Adams' p-validity [3, 4, 6, 9, 11]. P-validity allows to classify probabilistic inference rules as either "p-valid" or "p-nonvalid" [1] analog to "valid" and "nonvalid" rules in classical logic. Adams introduced p-validity in probability logic as a surrogate for validity in classical logic. P-validity functions as a substitute, an "Ersatz", when "... 'probable' and 'improbable' are substituted for 'true' and 'false'." [6, p.1] What's "validity" in classical logic is "p-validity" in probability logic.

Adams was not the first philosopher who emphasized the role of probabilities for reasoning. The outstanding pioneer was George Boole [20]. MacColl introduced the *suppositional* interpretation of conditionals and conditional probability. "The symbol $\frac{A}{B}$, when the numerator and denominator denote statements, expresses the chance that A is true on the assumption that B is true." [57, 58]. In the 1960s the development of modern probability logic started with Patrick Suppes [87]. He stimulated the beginning of the work of Ernest Adams [4]. Well-known became System P [52] and several of the closely related systems like the ϵ-semantic [65], System P^+ [81, 82] or System Z [65]. Some of these systems are "syntactically" equivalent to one of David Lewis' 27 systems investigated in the context of counterfactuals [55, 56].

In the psychology of reasoning p-validity was especially interesting since—when combined with the interpretation of the probability of conditionals as conditional probabilities, $P(\text{if } A \text{ then } B) = P(B|A)$—it classifies the probabilistic versions of some classically valid but counter-intuitive rules as p-nonvalid. The counter-intuitive PARADOXES OF THE MATERIAL IMPLICATION, CONTRAPOSITION, or STRENGTHENING THE ANTECEDENT are p-nonvalid. Moreover, the set of p-valid inference rules [4, p.277, Definition 6] corresponds to the rules of System P [36, 52]. Would human reasoning be closer to such a system than to a system of classical logic [83, 69, 71, 67]?

P-validity has been discussed in the psychological literature, for example, in [62, 47, 69]. Recently it has been claimed to be a "new standard" to evaluate the rationality of probability judgments in human reasoning [86, 31]. The present contribution throws a critical light on the application of p-validity in reasoning research. I first give an outline of the main arguments and topics of the contribution.

1.1 Preview

1. *Coherence:* Human judgments and inferences are evaluated by comparisons with "normative" standards. One such standard is *coherence* (de Finetti). A probability assessment is said to be coherent if it does not allow a Dutch Book. A Dutch Book is a bet where you lose for sure; you pay one Dollar for

[1] I follow Adams and write "p-nonvalid" and not "non-p-valid".

a bet in which you can maximally win only 99 Cents. The coherence approach to probability theory provides a *special* and (I think) fruitful perspective to model human uncertain reasoning. As Adams' p-validity is seen here on the background of the coherence approach I explain the concept of coherence in section 2.

2. *P-validity:* In classical logic an inference is valid if the conjunction of its premises implies its conclusion. In probability logic an inference is p-valid if the probability of the conjunction or the quasi-conjunction of its premises exceeds the probability of its conclusion. Quasi-conjunctions were introduced by Adams [6] to allow the conjunction of conditional events [39]. Adams expressed this criterion in terms of "uncertainties" where uncertainties are just the 1-complements of probabilities. The result is the uncertainty-sum criterion: The uncertainty of the conclusion may not exceed the sum of the uncertainties of the premises. P-validity is explained in section 3.

3. *Precise versus imprecise probabilities* P-validity refers to interval probabilities, not to point probabilities. Interval probabilities (or imprecise probabilities), assign lower and an upper bounds to each of the premises and infer an interval probability for the conclusion. P-validity refers to inferences where the probabilities of the premises are *higher* than a given value and less than or equal to 1. In psychological experiments human inferences were evaluated by the criterion of p-validity, but the participants in the experiments assessed point probabilities and not interval probabilities. As a consequence, incoherent judgments are claimed to be rational. We introduce the concept of p-validity in section 3.

4. *Generalized p-validity:* P-validity defines a threshold that depends upon the lower probabilities of the premises, that is, only upon the numerical values, not upon the logical form of the premises. The threshold is the same for any permutation of these values. Moreover, the uncertainty-sum criterion applies only to logically independent premises. Every-day reasoning may, however, require logically dependent premises. Section 4 introduces Adams' and Levine's [11] concept of essentialness leading to the so called generalized p-validity.

5. *Nonintuitive inference rules:* Classical logic contains rules which are valid but nonintuitive in everyday reasoning. Typical examples are the paradoxes of the material implication. In Adams' probability logic these rules turn out to be p-nonvalid. Do we need p-validity to protect against these nonintuitive rules? Pfeifer and Kleiter [70, 71, 74] have shown that the nonintuitive rules

are those which lead to conclusions with the vacuous probability interval $[0,1]$. We have called such rules "probabilistically noninformative" and argued that the intuitive rules are probabilistically informative, while the nonintuitive rules are probabilistically noninformative.

6. *Interpretation of conditionals:* Adams introduced "The Equation", which means to set the probability of an if-then statement equal to the probability of a conditional event. P-validity works in alliance with the conditional event interpretation of conditionals. The conditional event interpretation is clearly different from the material implication of classical propositional calculus. The interpretation of if-then sentences has extensively been studied in the psychological literature [32, 50, 68]. The relationship between p-validity and the interpretation of conditionals is discussed in section 6.

7. *Inconsistent premises:* P-validity presupposes premises with upper probabilities equal to 1. Not all sets of premises admit coherent upper bounds equal to 1 simultaneously for all premises. Adams called such cases "inconsistent premises". We will treat inconsistent premises in section 7. We will observe an interesting relationship to inferences in which if-then sentences are interpreted as material implications. The topic is treated in section 7.

8. *Inferring correlations:* From a psychological perspective it is often more important to consider inferences about correlations than inferences about propositions. Many psychological studies investigated the judgment of correlations. Section 9 treats inferences about 2×2 correlations in the context of probability logic and p-validity. From the premises of a MODUS PONENS, for example, we may infer lower and upper correlations.

9. *Zero probabilities of the conditioning event.* Kolmogorov's probability theory introduces conditional probability by the ratio-definition: $P(B|A) = P(A \wedge B)/P(A)$, where $P(A) \neq 0$. If $P(A) = 0$ conditional probabilities are undefined. Adams regrets "... the neglect of the possibility that the antecedents of the conditionals involved may have zero probability and we have no theory which applies to that case ... I arbitrarily stipulated that if $p(A) = 0$ then both $[p(B|A)]$ and $[p(\neg B|A)]$ equal 1 ..." [6, p.40, p. 46]. Setting $P(B|A) = 1$ if $P(A) = 0$ is called the *null-unity convention* [14]. While zero probabilities do not directly appear in the psychological literature, probabilities equal to 1 do [62, 63]. If we consider $P(B|A)$ and assume $P(A) = 1$ what about $P(B|\neg A)$? Zero probabilities are relevant in modeling uncertain reasoning [68] and are discussed in section 8.

10. *n-increasing probabilities:* Consider a probabilistic inference rule with n premises. Assume that if the probability of any of the n premises increases, then also the probability of the conclusion increases. Such rules may be called "n-increasing". Are p-valid rules n-increasing? No, the MODUS TOLLENS, for example, is not n-increasing but p-valid. In section 5 we speculate that human reasoning endorses n-increasing rules.

11. *Nothing is wrong with p-nonvalid rules:* Logical inference rules like DENYING THE ANTECEDENT or AFFIRMING THE CONSEQUENT are logically nonvalid. Their probabilistic versions are p-nonvalid. But why should the probabilistic versions of these rules be discredited? They propagate coherent probabilities of the premises to coherent probabilities of the conclusions corresponding to the rules of probability theory. Propagating probabilities coherently is just fine.

1.2 Preliminaries

A probabilistic inference rule has the form

$$\{(X_1, val_1), \ldots, (X_n, val_n)\} \models (Y, val_Y),$$

where

1. $\mathcal{X} = \{X_1, \ldots, X_n\}$ denotes a finite set of n premises. In basic inference rules the number of premises is just one or two. Each premise contains one or more events. The events are either unconditional or conditional events. Unconditional events are denoted by A, B etc. They are negated and combined like propositions in propositional logic. *If-then* sentences are either interpreted as *conditional events* or as *material implications.* Conditional events are denoted by $B|A$ (B given A, $A \neq \bot$). They have the truth or indicator values (i) 1 (TRUE) if both A and B have the values 1 (TRUE), (ii) 0 (FALSE) if A is 1 and B is 0, and (iii) the truth value is undetermined if A is 0 (FALSE). Conditional events are not propositions. They do not combine as conjunctions or disjunction in the usual way and they do not iterate in the usual way (but see [43], [9, p. 164]). Material implications are denoted by $A \to B$. They have the truth or indicator value 0 (FALSE) if A is TRUE and B is FALSE and the value 1 otherwise. $\mathcal{X}$ builds the logical carrier structure of the valuation of the premises. Adams calls the (truth-functional) language that contains only unconditional events *factual* and a language that contains also (non-truth-functional) conditionals its *extension.* An unconditional event may be represented by conditioning on the tautology; A is the short form of $A|\top$.

2. $val_1, \ldots, val_n$ is an associated valuation of the premises. The valuation may be a probability assessment; the assessment may be precise (point probabilities), imprecise (interval probabilities or second order density functions), or a mixture of both. In some cases, like in the LEFT LOGICAL EQUIVALENCE in System P, a premise is explicitly valuated by a truth value.

3. Y denotes the conclusion and val_Y its inferred valuation. Y contains again unconditional or conditional events. The valuation may be precise. However, in probability logic it is usually an interval probability.

4. $\models$ denotes the entailment relation. For finite numbers of premises p-validity is a *deduction relation*; it is monotone, transitive, and invariant with respect to the substitution of truth-functional formulas for atomic formulas (see Theorem 1.3 in [4, p. 275] or [9, p. 151]). For more details the reader is referred to chapter 3 in [84], especially to the definition of a "Wahrscheinlichkeitstheoretische Folgerung" in Def. 3-5. and to the 2012 paper of Schurz and Thorn [85].

We denote the lower and upper values of a probability interval by single or double inverted commas like $[\alpha', \alpha'']$, $0 \leq \alpha' \leq \alpha'' \leq 1$. In some sections we mark the probability of conditionals either by a vertical stroke "$|$", like $P(B|A) = \beta_|$, or by a right arrow "$\to$", like $P(A \to B) = \beta_\to$. The vertical stroke is used for the interpretation of conditionals as conditional events, the right arrow for the interpretation as material implications. Throughout I use "material implication" for material conditionals and "conditional events" for the conditional event interpretation of conditionals.

The inference rules most often investigated in the psychology of reasoning are the members of the quartet MODUS PONENS, MODUS TOLLENS, DENYING THE ANTECEDENT, and AFFIRMING THE CONSEQUENT (Table 1). The MODUS PONENS and

MODUS PONENS	MODUS TOLLENS	DENYING THE ANTECEDENT	AFFIRMING THE CONSEQUENT
A	$\neg B$	$\neg A$	B
$A \to B$	$A \to B$	$A \to B$	$A \to B$
B	$\neg A$	$\neg B$	A

Table 1: The four inference rules most often investigated in reasoning research.

the MODUS TOLLENS are valid, DENYING THE ANTECEDENT and AFFIRMING THE CONSEQUENT are nonvalid.

In probability logic the conditional "*if A then B*" is interpreted as a conditional event "$B|A$". Each premise is assigned a probability assessment which in the psychological experiments is usually precise, that is, a point and not an interval probability. The coherent probabilities of the conclusions are usually intervals. For the inference rules in Table 1 the corresponding probabilities are shown in Table 2.

MODUS PONENS	MODUS TOLLENS
$P(A) = \alpha$	$P(\neg B) = \alpha$
$P(B\|A) = \beta$	$P(B\|A) = \beta$
$P(B) \in [\alpha\beta, \alpha\beta + 1 - \alpha]$	$P(\neg A) \in [\max\{\frac{1-\alpha-\beta}{1-\beta}, \frac{\alpha+\beta-1}{\beta}\}, 1]$
DENYING THE ANTECEDENT	AFFIRMING THE CONSEQUENT
$P(\neg A) = \alpha$	$P(B) = \alpha$
$P(B\|A) = \beta$	$P(B\|A) = \beta$
$P(\neg B) \in [(1-\alpha)(1-\beta), 1-\beta(1-\alpha)]$	$P(A) \in [0, \min\{\frac{\alpha}{\beta}, \frac{1-\alpha}{1-\beta}\}]$

Table 2: The rules of Table 1 for precise probabilities of the premises and the inferred coherent lower and upper probabilities of the conclusion [71, p.212]. The bounds require A and B to be logically independent.

2 Coherence

One of the first publication of Adams' [2, 1, in two parts] shows that his work on probabilistic inference started with the investigation of rational betting systems, very much in the spirit of Ramsey and de Finetti and what today is called the *coherence approach* (compare also the Appendix 1 in [9]). Probabilities are introduced with the help of *conditional bets*, that is, by bets that A will be found true, given that B is true. Bets are called off if A will be found false. Adams was also highly familiar with the behavioral decision theory of the 1950ties. This is shown by work done together with Fagot on riskless choice behavior [10].

Coherence is one of the most fundamental concepts on which the foundation of de Finetti's theory of subjective probability in based. In philosophy and cognitive science *coherence* is a widely accepted rationality criterion of uncertain reasoning. In the behavioral sciences it functions as a normative benchmark to evaluate human inferences and judgments under uncertainty. Originally the concept was introduced for the assessment and the propagation of precise probabilities. It was extended by Walley [90, 91] to the paradigm of *imprecise probabilities* [12]. The imprecise probability approach assesses and propagates interval probabilities from premises to conclusions. Closely related is the work of Gilio [34, 38, 18, 19]. Of special interest is his study [36] on the relationships between System P, a nonmonotonic logical system [52] pioneered by Adams [6], and interval probabilities. Interval probabilities were extensively studied by Weichselberger [99, 96, 97].

The precise probability assessment of a sequence of conditional events is coherent if there exists no combination of bets which *certainly* results in a loss [24, 26, p. 87]: A *precise* probability assessment $\mathcal{P} = (\alpha_1, \ldots, \alpha_n)$ of a sequence of conditional events $\mathcal{X} = (B_1|A_1, \ldots, B_n|A_n)$ is coherent if and only if the random gain G given by

$$G = \sum_{i=1}^{n} \lambda_i |A_i| (|B_i| - \alpha_i) \tag{1}$$

is neither uniformly positive nor uniformly negative and if this holds for all subsets of $\mathcal{X}$; $|A_i| \in \{0, 1\}$ and $|B_i| \in \{0, 1\}$ denote the indicator values of A_i and B_i, and λ_i, denotes the "stakes" of a bet; λ_i, $-\infty < \lambda_i < +\infty$ may have any finite positive or negative real value. A bet with sure loss is often called a "Dutch Book". For unconditional events the factor $|A_i|$ simplifies to $|A_i| = |\top| = 1$.

Geometrically an interval assessment $[\alpha'_1, \alpha''_1], \ldots, [\alpha'_n, \alpha''_n]$ may be represented by an n-dimensional hyperrectangle (called an n-orthotope) resulting from the set of all points in the Cartesian product $[\alpha'_1, \alpha''_1] \times \cdots \times [\alpha'_n, \alpha''_n]$. An interval assessment $[\alpha'_1, \alpha''_1], \ldots, [\alpha'_n, \alpha''_n]$ is *g-coherent* (generalized coherence, [38, 40]) if its n-orthotope is (i) not empty and (ii) if *all* precisely coherent assessments $(\alpha_1 \in [\alpha'_1, \alpha''_2], \ldots, \alpha_n \in [\alpha'_n, \alpha''_n])$ are a subset of the n-orthotope. No coherently precise point $(\alpha_1, \ldots, \alpha_N)$ lies outside of the n-orthotope.

The left panel of Figure 1 shows a numerical example for a 3-orthotope. It represents the probabilities for a MODUS PONENS with the premises $P(A) = \alpha \in [.7, .9]$ and $P(B|A) = \beta \in [.4, .65]$, and the probability of the conclusion $(B) = \gamma \in [.28, .775]$. The lower and upper probabilities of the conclusion are $\gamma' = \alpha'\beta'$ and $\gamma'' = 1 - \alpha' + \alpha'\beta''$. The expressions are extensions of the formula for precise probabilities $P(B) \in [\alpha\beta, 1 - \alpha + \alpha\beta]$ (see Table 2). The interval assessment is g-coherent for the events A, $B|A$ and B of a MODUS PONENS with a conditional event interpretation of the conditional.

The right panel of Figure 1 shows an n-polytope. Here the lower and upper probabilities of the conclusion $P(B)$ are not constant but functions of the premise probabilities $P(A)$ and $P(B|A)$. We have again $\alpha \in [\alpha' = .7, \alpha'' = .9], \beta \in [\beta' = .4, \beta'' = .65]$, but now $\gamma \in [\alpha'\beta', 1 - \alpha' + \alpha'\beta'']$. Now *all* the points in the polytope are precisely coherent.

An interval assessment $[\alpha'_1, \alpha''_1], \ldots, [\alpha'_n, \alpha''_n]$ is *totally coherent* if all $(\alpha_1, \ldots, \alpha_n)$, $\alpha_1 \in [\alpha'_1, \alpha''_1], \ldots, \alpha_n \in [\alpha'_n, \alpha''_n]$, are precisely coherent. In Figure 1 the polytope on the right hand side is completely included in the 3-orthotope on the left hand side.

Computationally the coherence of a precise conditional probability assessment may be investigated by means of linear algebra. The indicator values of the atoms (constituents) generated by the events $(A_1|B_1, \ldots, A_n|B_n)$ are represented in a matrix $\mathbf{Q}$ and the assessed probabilities in a vector $\boldsymbol{\alpha}$. If the rank of $\mathbf{Q}$ and the rank of $\mathbf{Q}|\boldsymbol{\alpha}$, the matrix extended by the vector $\boldsymbol{\alpha}$, are equal, then the corresponding system of linear equations is solvable and the assessment is coherent. If the system is not solvable the assessment is not coherent.

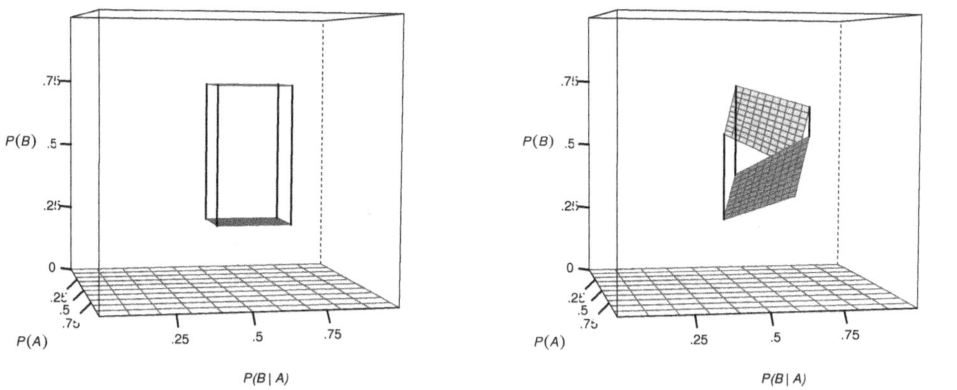

Figure 1: Left: A g-coherent interval assessment for a MODUS PONENS with $P(A) = \alpha \in [.7, .9]$, $P(B|A) = \beta \in [.4, .65]$ and $\gamma \in [.28, .775]$. Right: A totally coherent interval assessment for the same premises but *inferred* and not fixed probabilities of the conclusion, $P(B) = \gamma \in [\alpha\beta, 1 - \alpha + \alpha\beta]$. For fixed (α, β)-points (at the bottom) all points in the interval between the lower and upper surfaces are coherent.

For interval assessments different names for the same properties are used: *avoiding sure loss* in the framework of imprecise probabilities [90], *g-coherence* (generalized coherence) in the the framework of Gilio [35], or *R-probable* (reasonable) in the framework of Weichselberger [99, 96, 97, 98]. An interesting proposal involving second order probabilities was made by Bamber [13] and Bamber, Goodman, and Nguyen [14]. The second order probabilities may be used to determine the probability of being coherent (see also [49]).

The recent probabilistic paradigm on human reasoning tends to neglect premises with interval assessments. In the probability logic of Adams interval probabilities of the premises are, however, essential. Referring to Adams' p-validity means to refer to imprecise probabilities in the premises. As this is not always obvious in the literature let us give some more detailed explanation in the next section.

3 P-validity

Assume the premises of an inference consist of a sequence of events $X_1, \ldots, X_n$ together with the *interval* assessment $P(X_1) = \alpha_1 \in [\alpha'_1, 1], \ldots, P(X_n) = \alpha_n \in [\alpha'_n, 1]$. Assume that the lower and the upper bounds of the assessment are coherent. Following Adams, call the 1-complement of the probability of an event X_i its "uncertainty", $u(X_i) = 1 - P(X_i)$ [2]. Adams introduced the following *uncertainty-sum criterion* to define probabilistically valid (p-valid) inferences:

> The uncertainty of the conclusion of a [probabilistically] valid inference cannot exceed the sum of the uncertainties of its premises [9, p.38] [6, p.2].

Let us make this definition more explicit. The phrase "cannot exceed" refers to *two* upper bounds, one derived from the coherence and one derived from the uncertainty-sum criterion. Denote the upper *coherent* uncertainty of the conclusion Y by v''_Y and its upper *uncertainty-sum* by u''_Y, respectively. The definition requires that $u''_Y \leq v''_Y$. So we have:

Definition 1 (P-validity in terms of uncertainty). *Let R be a probabilistic inference rule with the form*

$$u(X_1) \in [0, u''_1], \ldots, u(X_n) \in [0, u''_n] \models u_Y \in [0, v''_Y], \quad 1 \leq n < \infty .$$

Assume the premises $X_1, \ldots, X_n$ to be logically independent. v''_Y denotes the upper coherent uncertainty of the conclusion Y. Let u''_Y denote the uncertainty-sum criterion given by

$$u''_Y = \sum_{i=1}^{n} u''_i . \tag{2}$$

R is p-valid if for all coherent assessments of the $(\alpha'_1, \ldots, \alpha'_n)$ it holds that $v''_Y \leq u''_Y$.

As without further restrictions u''_Y might be greater than 1, we replace Equation (2) by Equation (3)

$$u''_Y = \min \left\{ 1, \sum_{i=1}^{n} u''_i \right\} . \tag{3}$$

[2] In his 1975 book [6, p.2 and 41] and in [11, p.429] Adams gives the warning: "... (where uncertainty is here defined as probability of falsity - not to be confused with the entropic uncertainty measure of Information Theory)." Despite this warning, in a recent paper the authors [47, p.3] identify Adams' uncertainty by means of Shannon's entropy.

Adams' uncertainties are 1-complements of probabilities and his phrase "... the uncertainty cannot exceed ..." refers to an upper bound of an interval, the lower bound of which is zero by default. The number of premises is finite ([6, p.52]). As it is more common to work with probabilities and not with 1-complements of probabilities we rephrase Definition 1 in terms of probability (compare also [36]).

Definition 2 (P-validity in terms of probability). *Let R be a probabilistic inference rule with the form*

$$P(X_1) \in [\alpha'_1, 1], \ldots, P(X_n) \in [\alpha'_n, 1] \models P(Y) \in [\theta', 1], \ 1 \leq n < \infty.$$

Assume the premises $X_1, \ldots, X_n$ to be logically independent. θ' denotes the lower coherent probability of Y. Let γ' denote the lower probability corresponding to the uncertainty-sum criterion

$$\gamma' = \left[\max\left\{0, \sum_{i=1}^{n} \alpha'_i - (n-1)\right\}, 1\right]. \tag{4}$$

R is p-valid if for all coherent assessments of the $(\alpha'_1, \ldots, \alpha'_n)$ it holds that $\gamma' \leq \theta'$.[3]

Figure 2 shows a numerical example of the uncertainties and the associated probabilities for the uncertainty-sum criterion applied to the MODUS PONENS. The inequality $\gamma' \leq \theta'$ in Definition 2 states that the lower p-validity bound is smaller than the lower coherence bound. The values in the interval $[\gamma', \theta']$ are incoherent, those in $[\theta', 1]$ are coherent. The uncertainty-sum criterion is g-coherent and *avoids sure loss*.

In 1966 Suppes [87] proved that an inference from a finite set of unconditional events to an unconditional event is p-valid exactly if it is classically valid (for a proof see also [84, p. 31 and p. 191]). Patrick Suppes was the supervisor of the PhD thesis of Ernest Adams.[4]

Example 1 (MODUS PONENS). *The coherent lower and upper probabilities of the conclusion of a* MODUS PONENS *(see Table 1 and Table 2) may be obtained with the Theorem of Total Probability $P(B) = P(A)P(B|A) + P(\neg A)P(B|\neg A)$ (see Figure 3). Let $P(A) = \alpha \in [\alpha', 1]$ and $P(B|A) = \beta \in [\beta', 1]$. In terms of probabilities the uncertainty-sum criterion leads to $\gamma' \in [\max\{0, \alpha' + \beta' - 1\}, 1]$. As $P(B|\neg A)$ is not given it may have any value between 0 and 1. If $P(B|\neg A) = 0$ we have*

[3]Questions about upper probabilities $\gamma'' = 1$ arising from zero probabilities of conditioning events will be discussed below in Section 8.

[4]The thesis, however, was written ten years earlier and on rigid body mechanics, a field rather different from probability logic.

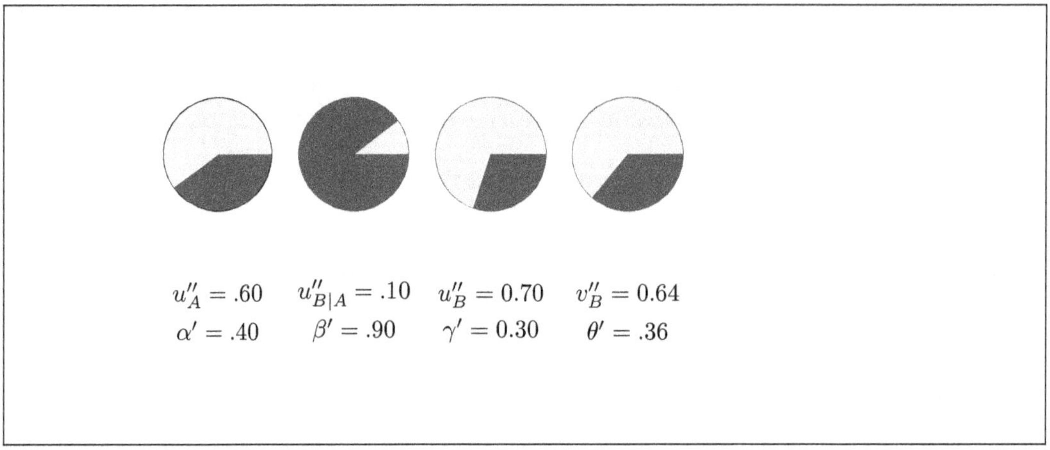

Figure 2: 1-complements of upper uncertainties (yellow on the web) and lower probabilities (blue on the web) for a MODUS PONENS with $u_A \in [0, .60]$ and $u_{B|A} \in [0, .10]$. The uncertainty-sum criterion is $u''_B = .10 + .60 = .70$ or in terms of lower probability $\gamma' = .30$. The lower probability of the MODUS PONENS is the product $\theta' = \alpha'\beta' = .36$. As for all α and $\beta \in [0, 1]$ $\gamma' < \theta'$ the MODUS PONENS is p-valid.

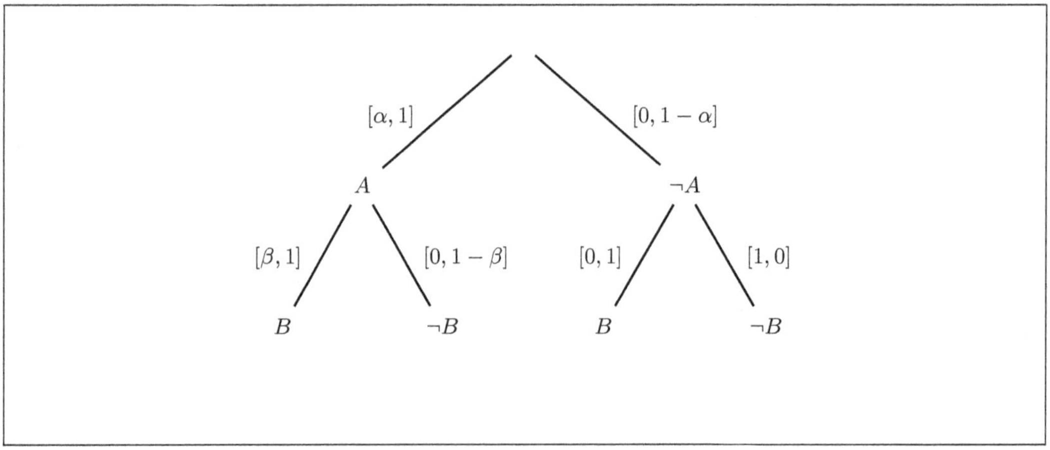

Figure 3: MODUS PONENS: Inferring the lower and upper probabilities of B, $P(B) \in [\alpha'\beta', 1]$, from an interval assessment of the premises $P(A) \in [\alpha', 1]$ and $P(B|A) \in [\beta', 1]$. As $P(B|\neg A)$ is not given it may have any value in $[0, 1]$.

$P(B) = P(A)P(B|A)$ so that the coherent solution is $\theta' \in [\alpha'\beta', 1]$. As $\gamma' \leq \theta'$, $0 \leq \alpha, \beta, \leq 1$, the MODUS PONENS *with conditional event interpretation is p-valid.*[5] *We will treat the question of a conditioning event with zero probability $P(A) = 0$ below.*

P-validity is a property that characterizes inferences. It may also be used to classify individual probability judgments as falling in- or outside of a "p-validity interval". If, however, in a psychological experiment the participants assess point probabilities for the premises and the probability judgments of the conclusion are evaluated by the uncertainty-sum criterion, then incoherent judgments are classified as "p-valid" or "rational".

The following example shows that for an important special case, the conjunction, the "p-validity interval" and the "coherence interval" of the conclusion are identical.

Example 2 (AND-INTRODUCTION). *From $P(A_1) \in [\alpha'_1, 1]$ and $P(A_2) \in [\alpha'_2, 1]$ we infer $P(A_1 \wedge A_2) \in [\max\{0, \alpha'_1 + \alpha'_2 - 1\}, 1]$. In this case the uncertainty-sum criterion γ' and the coherent lower probability θ' are identical.* AND-INTRODUCTION *is p-valid. Figure 4 shows the three-dimensional representation of the lower and upper probabilities and the linear contour lines for the lower probability of the conclusion.*

The uncertainty-sum criterion is nothing else than the lower probability of the conjunction *of the premises.*

While for simple problems lower and upper probabilities may be obtained by Gaussian elimination, more complex problems, are solved by linear programming. We arrange the indicators in an $(n+1) \times m$ matrix of coefficients $\mathbf{Q}$; n denotes the number of premises; one additional row is added representing the sure event Ω. m denotes the number of constituents. For r logically independent events $m = 2^r$. We denote the vector of the m indicators of the conclusion by $\mathbf{c}^T$ and the probabilities of the constituents by $\mathbf{x} = \begin{pmatrix} x_1 \\ \dots \\ x_m \end{pmatrix}$. The objective function of the linear program corresponds to the probability of the conclusion, that is, to $\mathbf{c}^T\mathbf{x}$. The linear program finds the lower probability

$$\min \mathbf{c}^T\mathbf{x} \quad \text{under the constraints} \quad \mathbf{Q}\mathbf{x} = \mathbf{b}, x_i \geq 0, i = 1, \dots, m. \qquad (5)$$

[5]One of the reviewers points out that Adams' uncertainty-sum criterion does not always lead to the tightest interval bounds because the purpose of Adams' notion of p-validity and p-entailment was to give a complete semantics for the p-calculus: p-validity is derivation-independent and holds for *all* inferences that can be derived from the p-calculus! The price for this is that p-valid intervals are not always tight.

Figure 4: Left: Three-dimensional representation of the lower probability of the CONJUNCTION, $\max\{0, P(A)+P(B)-1\}$ and the upper bound 1. Right: Contour lines.

and the upper probability by maximizing the objective function. For the conjunction we have

$$\mathbf{Q} = \begin{pmatrix} 1 & 1 & 0 & 0 \\ 1 & 0 & 1 & 0 \\ 1 & 1 & 1 & 1 \end{pmatrix}.$$

The first two rows correspond to the indicators of A and B, respectively. The third row represents the sure event Ω. The indicators of the conjunction are $\mathbf{c}^T = (1, 0, 0, 0)$. The linear program finds the minimum and the maximum of $\mathbf{c}^T\mathbf{x}$.

In the case of precise premise probabilities $\mathbf{Qx} = \mathbf{b}$ is a system of linear equations. In the case of imprecise premises the probabilities $\mathbf{b}$ are replaced by the lower probabilities of the premises and the "=" are replaced by "≥"relations so that a system of linear inequalities $\mathbf{Qx} \geq \mathbf{b}$ results. It checks for the existence of a coherent solution for the weaker constraints.

If the premises contain conditional events the indicator values in the coefficient matrix $\mathbf{Q}$ are replaced by the values of the conditional probabilities in those cases in which the indicator of conditioning event is zero. In the case of the MODUS PONENS, for example, we have

$$\mathbf{Q} = \begin{pmatrix} 1 & 1 & 0 & 0 \\ 1 & 0 & \beta & \beta \\ 1 & 1 & 1 & 1 \end{pmatrix}.$$

β is the conditional probability $P(B|A)$. It denotes the "money back" condition of the conditional bet on B if the conditioning event A is false. In the case in which the

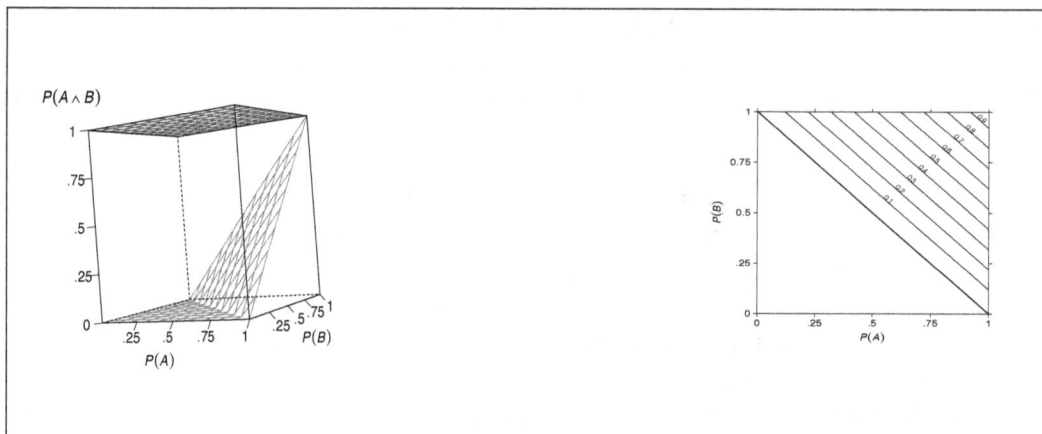

conclusion is a conditional event the problem is solved by fractional programming. Its application to probabilistic inference is explained in [54]. Numerical solutions of linear programming problems may easily be obtained with the help of Matlab [59] or R [79].

When referring to the uncertainty-sum criterion, the literature—not only the psychological literature—is often silent about the upper probability bounds of the premises. In Adams' framework they are—often *implicitly*—assumed to be 1. Let us repeat the uncertainty-sum criterion: "if an inference is truth-conditionally sound then the uncertainty of its conclusion cannot exceed the sum of the uncertainties of its premises ...". In terms of uncertainty this means that the premises may have any value between 0 and the uncertainty-sum value. In terms of probabilities this means that the probability of the premises may be 1. In psychologically relevant situations this is unrealistic. It may lead to study the propagation of overconfidence [62, 86, 31]. Also in well-known philosophical references such as [17, p. 131] it is not clear that the premises are interval probabilities. The uncertainty-sum criterion lures the understanding that the uncertainties are 1-complements of point probabilities while, corresponding to Adams' p-validity, they are 1-complements of *bounds* of *interval probabilities*. What is called the "uncertainty" of the conclusions is actually the upper bound of an interval with a lower bound equal to zero, that is, $[0, u''_B]$.

4 Generalized p-validity

The uncertainty-sum criterion is insensitive to the logical form of an inference. Whatever the logical interdependence of the premises, for the same numerical values of their probabilities the uncertainty-sum is the same. Adams met the insensitivity to the logical structure in two ways: (i) he generalized the p-validity criterion so that it becomes sensitive to the *essentialness* of the premises and (ii) he investigated lower *coherent* probabilities (sometimes called "worst case probabilities" by Adams). We turn to the generalized p-validity first and discuss the lower coherent probabilities in the section on p-entailment and probability preservation below.

The generalized p-validity criterion was introduced by Adams and Levine [11],[9, p.41ff.], [64, p.15]. It incorporates the inferential *essentialness* of the premises with respect to the specific conclusion at hand.

A premise A_i is essential with degree $e_i = 1$ if its removal from the set of premises makes the inference logically nonvalid. If a premise A_i is not a member of any essential subset of the premises, then its degree of essentialness e_i is 0. Otherwise A_i belongs to one or more sets of essential premises. If the cardinality of the set with the smallest number of premises to which A_i belongs is k, then $e_i = 1/k$. Now we have the following generalized p-validity criterion:

Definition 3 (Generalized p-validity). *Let R be a probabilistic inference rule as in Definition 2. Let θ' denote the coherent lower probability of the conclusion Y. Let γ' denote the lower probability corresponding to the p-validity criterion, now given by*

$$\gamma' = \max\left\{0, 1 + \sum_{i=1}^{n} e_i \alpha'_i - \sum_{i=1}^{n} e_i\right\}. \tag{6}$$

R is p-valid if for all coherent assessments of the α'_i it holds that $\gamma' \leq \theta'$.

Compare [9, p. 44] or Theorem 3.5 in [64]. Equation (6) builds weighted averages of the probabilities of premises in essential sets with identical cardinalities. Definition 2 refers to the special case in which $e_1 = 1, \ldots, e_n = 1$.

Only elementary rules such as the MODUS PONENS have premises with all $e_i = 1$. In these cases the uncertainty-sum criterion of Definitions 1 and 2 require no "correction". Equation (6) reduces to Equation (2). In many every-day inferences, however, the especially simple version of the uncertainty-sum criterion is not applicable as the following example shows.

Example 3 (OR-INTRODUCTION). *Adams [9] gives the example*

From $\{P(A) = .9, P(B) = .9, P(C) = .9\}$ infer $P(AB \vee AC \vee BC) \in [.85, 1]$.

(Writing AB for $A \wedge B$ etc.) None of the individual premises is essential but each one belongs to a set with two members, thus $e_1 = e_2 = e_3 = 1/2$

	x_1	x_2	x_3	x_4	x_5	x_6	x_7	x_8	α'_i
A	1	1	1	1	0	0	0	0	.9
B	1	1	0	0	1	1	0	0	.9
C	1	0	1	0	1	0	1	0	.9
$AB \vee AC \vee BC$	1	1	1	0	1	0	0	0	$[.85, 1]$
$\neg(AB \vee AC \vee BC)$	0	0	0	1	0	1	1	1	$[0, .15]$

Table 3: Inference with premises having essentialness $k = 1/2$ (Adams [9, p.42]). AB denotes the conjunction $A \wedge B$. The last row shows the conjugacy property.

With Equation (6) we obtain Adams' lower bound $\gamma' = 1 + 3\ \frac{1}{2}\ .9 - 1.5 = .85$.

With the help of Table 3 we have the linear system

$$\begin{aligned} x_1 + x_2 + x_3 + x_4 &= .9 \\ x_1 + x_2 + x_5 + x_6 &= .9 \\ x_1 + x_3 + x_5 + x_7 &= .9 \\ \sum_{i=1}^{8} x_i = 1, &\qquad x_i \geq 0, i = 1, \dots, 8\,. \end{aligned}$$

and the objective function $x_1 + x_2 + x_3 + x_5 = \gamma$. The lower bound of the disjunction is $\gamma' = .85$. Applying the uncertainty-sum formula of Definition 2 would lead to the value .7. Adams' lower probability is identical with the coherence solution, $\gamma' = \theta'$. Moreover, the upper probability of the conclusion is 1. But this is not generally true. If $P(A) = P(B) = P(C) = .6$, for example, we obtain $\gamma \in [.4, .9]$. Adams is misleading when he seems to assign point probabilities to the premises: "Suppose also that each premise has probability .9 and uncertainty .1." [9, p.42].

In the example all premises have the same probability. For premises with different probability assessments the generalized p-validity bounds are not the tightest ones [9, Footnote 7, p. 45]. We come back to this condition below.

Essentialness has not been discussed in the psychological literature and also some logical references do not mention it [17]. It is important even for such elementary rules as OR-INTRODUCTION.

Consider a committee with n members—think, for example, of the five permanent members of the UN Security Council. The members vote for or against a resolution. The probability that the ith member votes in favor of the resolution is α_i. What is the probability that exactly r members vote in favor of the resolution? If the votes are assumed to be independent and the probabilities are all equal, $\alpha_i = \alpha$, $i = 1, \dots, n$, then the probability of r successes in n trials is given by the binomial $P(r|n, \alpha, \text{ind}) = \binom{n}{r}\alpha^r(1-\alpha)^{n-r}$. Results for the case in which the independence assumption is dropped and for exchangeable events are given in [93, 26, 48].

What is the probability of r *or more* votes in favor of the resolution? In the case of independence it would be the sum of the binomial probabilities of $r, r+1, \dots, n$ successes. If we do not assume independence and if all probabilities are equal, then the lower coherent probability of r or more successes is

$$\theta'(s \geq r|n, \alpha) = \frac{\max\{0, n\alpha - (r-1)\}}{n - r + 1}\,. \tag{7}$$

For $r = n$ events this is the lower probability of the conjunction of r events. As n increases and r is kept constant, also $n - r + 1$, the number of the events outside of the conjunction increases; the lower conjunction probability is allocated to more and more events and gets smaller and smaller. Equation (7) is equivalent to Equation (14) in Adams and Levine [11, p. 446] if probabilities are replaced by their 1-complements (uncertainties) and "$\geq$" is replaced by "$>$". It may easily be seen that Equation (7) and Equation (6) are identical:

$$\begin{aligned} 1+\sum_{i=1}^{n} e_i\alpha - \sum_{i=1}^{n} e_i &= 1+\frac{n\alpha}{n-r+1} - \frac{n}{n-r+1} \\ &= \frac{n\alpha-(r-1)}{n-r+1} \end{aligned}$$

For the disjunction of r out of n events the degree of essentialness of the events is $e_i = 1/(n-r+1)$, $i=1,\ldots,n$, for $r>0$ and 0 otherwise.

If the probabilities $\alpha_i, i=1,\ldots,n$, are not all equal the essentialness criterion "approximates" the coherent solution by setting all probabilities equal to their mean. This leads to a bound that is *below* the coherent bound. Adams remarks in a footnote [9, p. 44/45] that his Theorem 15∗ "... is not the worst case uncertainty of its conclusion ...".

Let us briefly show how the "approximation" can be replaced by an exact value. The coherent lower bound for the case in which the probabilities are not equal and the events are not independent is obtained as follows. (i) We order the probabilities in descending order, $\alpha_1 \geq \alpha_2, \ldots, \alpha_{n-1} \geq \alpha_n$. (ii) We build a sequence of $w_k, k = r, \ldots, n$

$$w_k = w_{k-1} + \frac{\alpha_k - w_{k-1}}{k-r+1}, k = r+1,\ldots,n \tag{8}$$

starting with the conjunction probability of the most probable r events

$$w_r = \max\left\{0, \sum_{k=1}^{r} \alpha_k - (r-1)\right\} \tag{9}$$

and stop the sequence at k if the next ratio $\frac{\alpha_{k+1}-w_k}{k-r+1}$ is negative. The lower probability is

$$\theta'(s \geq r|n, \alpha_1, \ldots, \alpha_n) = w_k, \quad \text{where} \quad w_r, \ldots, w_k > 0 \quad \text{and} \quad w_{k+1} \leq 0 \tag{10}$$

(iii) If $k=n$, i.e., all ratios are positive, the lower probability is

$$\theta'(s \geq r|n, \alpha_1, \ldots, \alpha_n) = \frac{\max\{0, \sum_{k=1}^{n} \alpha_k - (r-1)\}}{n-r+1}, \tag{11}$$

if $k<n$

$$\theta'(s \geq r|n, \alpha_1, \ldots, \alpha_n) = w_k\,. \tag{12}$$

The start value w_r is the same as in the case in which all probabilities are equal. The $w_r, \ldots, w_k$ iteratively tighten the interval of the probability of the conclusion.

Example 4 (Non-equal probabilities). *Let $A_1, \ldots, A_5$ be $n = 5$ events having probabilities $\alpha_1, \ldots, \alpha_5 = (.80, .75, .70, .50, .10)$. What is the lower probability of at least $r = 2$ successes? Corresponding to the Adams and Levine approach the essentialness of each of the 5 events is $e_i = 1/(n - r + 1) = 1/4$ and the lower probability of 2 or more successes is $1 + \sum_{i=1}^{5} e_i\alpha_i - \sum_{i1}^{n} e_i = 0.4625$. The mean of the five α_i is .57. Replacing the probabilities $(.80, .75, .70, .50, .10)$ by their mean $(.57, .57, .57, .57, .57)$ leads to the same result .4625.*

We next determine the lower probability corresponding to the coherence criterion. The probabilities are already ordered descendingly. The lower conjunction probability of the first $r = 2$ events is $.80 + .75 - 1 = .55$ so that $w_2 = .55$; now w_3 is $w_3 = w_2 + (\alpha_3 - w_2)/(3 - 2 + 1) = .55 + (.70 - .55)/2 = .625$. In the next step $w_4 = w_3 + (\alpha_4 - w_3)/(4 - 2 + 1)$, the ratio $(.5 - .625)/3 = -.0417$ is negative so we stop here and $\theta'(s \geq 2|n = 5, \alpha_1 = .8, \alpha_2 = .75, \alpha_3 = .7, \alpha_4 = .5, \alpha_5 = .1) = .625$. The coherence bound leads to the tightest lower bound which, in the example, is much higher than the bound resulting from the essentialness criterion.

To summarize: In the case of equal probabilities the essentialness result and the coherence result coincide. In the case of non-equal probabilities the essentialness criterion results in a g-coherent "approximation" of the coherent result. The essentialness criterion applies to factual formulas only, not to "... inferences involving *conditional* propositions, whose probabilities are plausibly measured as conditional probabilities." [emphasized in the original text] [11, p.431]

5 P-entailment and probability preservation

Do probabilistically reasonable inference rules always lead from highly probable premises to highly probable conclusions? Is the probability of the conclusions increasing if the probability of any of its n premises is increasing, i.e., is it n-increasing? Adams introduced two properties that investigate these questions, *probabilistic entailment* (p-entailment) and *high probability preservation*. Let us turn to p-entailment first.

Definition 4 (P-entailment). *Let R be an inference rule with the form*

$$P(X_1) \in [t, 1), \ldots, P(X_n) \in [t, 1) \models P(Y) \in [\theta', 1)\,.$$

R is said to probabilistically entail (p-entail) its conclusion if for all coherent probability assessments of the premises it holds that

$$\forall\theta' \in (0,1)\ \exists t \in (0,1)\ \Big((P(X_1) \in [t,1), \ldots, P(X_n) \in [t,1)) \models P(Y) \in [\theta',1)\Big). \quad (13)$$

If we fix the lower probability of the conclusion then there exists a threshold t such that all probabilities of the premises that are higher than the threshold guarantee the minimum probability of the conclusion; t is chosen such that it is the coherent minimum probability to infer the conclusion with lower probability θ'. In the definition the backward direction is important, from the conclusion to the premises. We note that the intervals are semiopen, that is, probability 1 is not included.

P-entailment was introduced by Adams in 1966 and originally called *reasonable consequence* [4, p. 274]. It was discussed in many of his later contributions [11, 6, 9]. Following Pearl [65, 66] it is often called ε-entailment. P-entailment does not require to exclude conditioning events with probability zero. As one of the reviewers points out Adams [6, 5, 7] distinguishes ϵ- and p-entailment (see also [85]); ϵ-entailment restricts the probability assignments to those which are proper while p-entailment does *not*.

For various inference rules Figure 5 shows on the X-axis the lower probability of the conclusion, θ', and on the Y-axis the threshold t for the premises required to guarantee the probability θ' of the conclusion. The left part shows inference rules which are p-entailing, the right part shows rules that are not p-entailing. The thresholds are obtained by setting all lower probabilities of the conclusion to θ' and solving the corresponding formulas for t. For the conjunction, e.g., we obtain from $\theta' = \alpha_1 + \alpha_2 - 1$ the value $t = (1 + \theta')/2$. The lower bound $t = 1/(2 - \theta')$ is the same for IF-THEN-INTRODUCTION, MODUS TOLLENS, and CAUTIOUS MONOTONICITY (from $\{P(B|A), P(C|A)\}$ infer $P(C|A \wedge B)$).

Benferhart, Dubois, and Prade [16] have shown that p-entailment also holds for non-infinitesimal probabilities; more specifically, it holds for $P(A \wedge B) > P(A \wedge \neg B)$ or $t > .5$. It is interesting to note that [29] and [30] put the comparison of the two conjunctions at the heart of the psychological interpretation of the Ramsey test. The $t = .5$-threshold is, of course, a psychologically highly plausible and designated value.

When the probability of a conditional event approaches 1, then its probability will get closer and closer to the probability of a material implication. Suppes [88, p.10] took the MODUS PONENS as an example. If $P(A) = 1 - \varepsilon$ and $P(B|A) = 1 - \varepsilon$ then the lower bound for $P(B)$ is $(1 - \varepsilon)^2$. For the material implication $P(A \to B)$ the result is $1 - 2\varepsilon$. For very small ε the difference is of little practical value. "Surely such a small difference in itself cannot have much psychological significance ..." [88].

What about premises with probability 1? The question leads to what Adams called *strict validity* in 1966 [4, Definition 5.2, p. 274] and *weak validity* in his 1998 book [9, p. 152]: An inference form is *weakly valid* if: if all its premises are certain then its conclusion is certain [9, p. 152]. In other words, if the inference form is certainty-preserving.

Adams takes, as an example, one of the paradoxes of the material implication, from "A" infer "if $\neg A$ then B". "... if the probability of A is not only close to but actually equal to 1, then the probability of $B|\neg A$ must also equal 1" [9, p. 152, changed notation]. If $P(A) = 1$, then the probability of the conditioning event must be zero, $P(\neg A) = 0$. In this case, corresponding to the Kolmogorov approach the conditional probability is not defined. Adams' way out [9, p. 152, footnote 5] is to take $P(B|\neg A) = 1$. Following this proposal makes, however, the paradox of the material implication strictly valid! The paradoxes of the material implication are, however, nonintuitive and empirical investigations have shown that humans do not endorse them [74]. Conditional probability logic is psychologically attractive because it keeps intuitive inference rules and discards nonintuitive ones such as the paradoxes of the material implication.

The coherence approach to probability avoids the difficulty. If $P(A) = 1$ and therefore $P(\neg A) = 0$ then all we can say about $P(B|\neg A)$ is that it is in the vacuous interval $[0, 1]$, i.e., that the inference is probabilistically noninformative [70, 74].

Example 5 (DENYING THE ANTECEDENT). *To infer from "$\neg A$" and "if A then B" the conclusion "$\neg B$" is logically nonvalid. In terms of probabilities, we remember from Table 2 that if $P(\neg A) = \alpha$ and $P(B|A) = \beta$, then $P(\neg B) \in [(1-\alpha)(1-\beta), 1-\beta(1-\alpha)]$. We see that if $\alpha = \beta = t$ and thus $\theta' = (1-t)^2$, θ' is a monotonically* decreasing *function of t.*

Let us fix the lower probability of $P(\neg B)$ at $\theta' = .5$; this requires a t-value of .293. The existence of just one t for which $\theta' = .5$ is satisfied is not enough; $\theta' = .5$ must be satisfied for all $t \geq .293$ *and this is clearly not the case.* DENYING THE ANTECEDENT *does not p-entail its conclusion.*

The example shows that in the definition of p-entailment the all-quantifier is crucial. As a consequence, psychological experiments which investigate p-entailment cannot do that with judgments about single numerical examples. Such experiments would require to keep values of θ' fixed and vary the values of t.

Let us remark that following Adams condition of proper (non-zero) probability assignments we specify $P(\neg A)$ by the semiopen interval $[t, 1)$ to avoid a zero probability for A in $P(B|A)$. In the coherence approach this is not necessary. Here $P(B|A) \in [0, 1]$ for $P(A) = 0$, i.e., the inference is probabilistically noninformative.

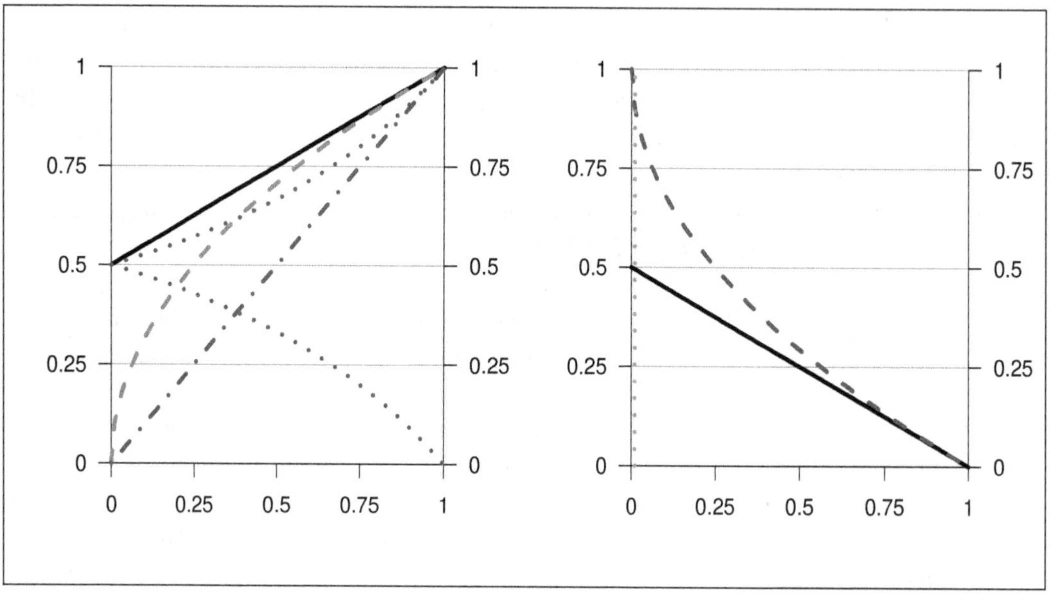

Figure 5: Lower probability of the conclusion θ' on the X-axis, threshold t for the lower probabilities of the premises on the Y-axis. P-entailment (left): AND-INTRODUCTION (solid line), MODUS PONENS (dashed), MODUS TOLLENS (dotted), and OR-INTRODUCTION (dotdash). Non-p-entailment (right): NON-AND-INTRODUCTION (solid line), DENYING THE ANTECEDENT (dashed), AFFIRMING THE CONSEQUENT (dotted at $\theta' = 0$).

P-validity and p-entailment are equivalent (First Equivalence Theorem in [9, p. 152]). We next consider, however, an example which is p-valid, p-entailing but not n-increasing.

Example 6 (MODUS TOLLENS). *Setting α and β equal to t in the formula for the* MODUS TOLLENS *in Table (2)*

$$\{P(B|A) = t, P(\neg B) = t\} \models P(\neg A) = \theta \in \left[\max\left\{\frac{1-2t}{1-t}, \frac{2t-1}{t}\right\}, 1\right] \tag{14}$$

and solving for t gives (i) $t \geq \frac{1-\theta'}{2-\theta'}$ if $t \leq .5$ and (ii) $t = \frac{1}{2-\theta'}$ if $t \geq .5$. Wagner [89, p. 752] derives the lower and upper bounds and for $\varepsilon \leq .5$ (corresponding to our $1-t > .5$) compares the lower bound with Suppes' ε-semantics. For various inference rules Figure (5) shows the relationship between the lower probability of the conclusion (θ', on the abscissa) and the minimum probabilities of the premises (t, on the ordinate) required to guarantee θ'. The MODUS TOLLENS *requires either highly probable* *(increasing dotted line) or highly low probable (decreasing dotted line) premises to generate a highly probable concluson. Figure 6 compares the* MODUS PONENS *and the* MODUS TOLLENS *in a three-dimensional representation. The "V-shaped" representation of the* MODUS TOLLENS *differs from the monotonically increasing shape of the* MODUS PONENS. *Figure (7) shows the corresponding contour lines. For* $\alpha = \beta = .5$ *the lower coherent probability of the* MODUS TOLLENS *is* $\theta' = 0$. *As for* $\alpha = \beta = .5$ *also the uncertainty-sum criterion is* $\gamma' = 0$ *the p-validity criterion is satisfied. The* MODUS TOLLENS *is not n-increasing, but it is p-entailing. For every* θ' *there exists a threshold t such that Definition 4 is satisfied. Adams did not fully analyze the* MODUS TOLLENS. *He writes [6, Note 2, p. 68] " ... the maximum uncertainty of the conclusion* $\neg A$ *of the ... Modus Tollens inference with premises* $B|A$ *and* $\neg B$ *equals the uncertainty of* $\neg B$ *divided by the probability of* $B|A$*." [changed notation, italic in the original text] This means that* $u'' = (1 - \alpha)/\beta$, *but this holds only for* $\alpha + \beta > 1$, *but not for* $\alpha + \beta < 1$.

Figure (5) and (6) compare some of the most elementary inference rules. Usually people endorse those rules for which the probabilities of the premises and the conclusion are proportional. The MODUS PONENS is "easy" while the MODUS TOLLENS is "difficult". The relationships between the probabilities of the premises and the conclusions may empirically be investigated by calculating the correlations between the corresponding probability judgments.

We call a rule in which the probability of the conclusion increases monotonically with the probability of each of its n premises "n-increasing". AND-INTRODUCTION and the MODUS PONENS are 2-increasing, but the MONDUS TOLLENS is not 2-increasing. This can easily be seen in the three-dimensional representations in Figure 6 and the corresponding contour lines in Figure 7. Copula-functions are n-increasing. The lower bound of the MODUS PONENS is in fact a copula (the product copula), the lower probability of the MODUS TOLLENS, however, is not a copula. Copulas may be ordered by a dominance relation inducing a partial order, i.e., an order which is reflexive, antisymmetrical and transitive, but in which not necessarily all members are comparable. On the left side of Figure 5 lower curves probabilistically dominate non-crossing upper curves, OR-INTRODUCTION $\succ$ MODUS PONENS $\succ$ AND-INTRODUCTION, while the MODUS TOLLENS is non-comparable. The dominance relation is an indicator of the strength of an inference rule.

Evans et al. [31, Table 3] use the uncertainty-sum criterion $\alpha + \beta - 1$ for the lower and 1 for the upper probabilities of DENYING THE ANTECEDENT and AFFIRMING THE CONSEQUENT. The uncertainty-sum criterion is, however, not applicable because both inference rules are not p-valid. Incoherent probabilities are systematically classified as endorsing the DENYING THE ANTECEDENT. The upper probability 1

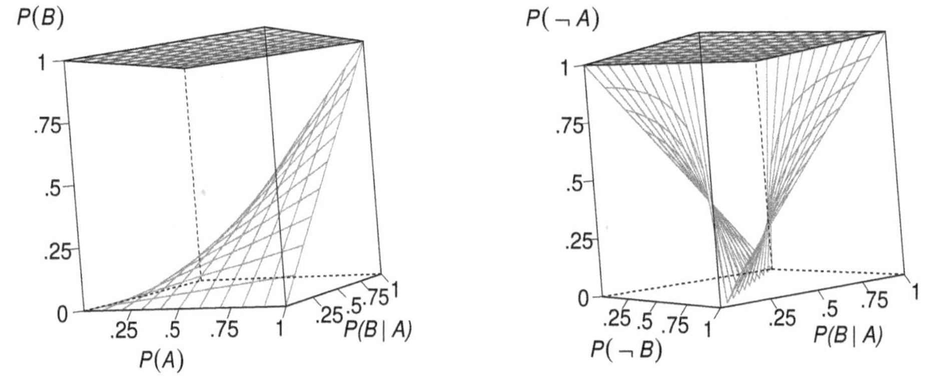

Figure 6: Left: MODUS PONENS. Right: MODUS TOLLENS. Lower and upper probabilities inferred from interval probabilities of the premises $P(A)$ ($P(\neg B)$) and $P(B|A)$ with upper bounds 1. On the left hand side the probability of the conclusion is a monotonically increasing function of the probabilities of the premises, on the right hand side it is not.

would be fine if the probability assessments of the premises would be interval probabilities $\alpha \in [\alpha', 1]$ and $\beta \in [\beta', 1]$. But the participants of the experiment assessed point probabilities so that the coherent upper probabilities are less than 1.

In addition to p-validity and p-entailment Adams introduced a hierarchy of four probability-preservations [8]. Of these *high probability preservation* is most relevant in the present context. It removes the difficulty with low probabilities in the MODUS TOLLENS.

An unconditional event A is *highly probable* if and only if the highest probability of those constituents where A is true is higher than the highest probability of those constituents where A is false [9, p. 133 ff.]. A conditional event $B|A$ is *highly probable* if and only if the highest probability of those constituents where the conjunction $A \wedge B$ holds is higher than the highest probability of those constituents where $A \wedge \neg B$ holds.

To find out whether an event is *highly probable* or not, we rank order the constituents by their probabilities. If the highest rank in the set of constituents where $A = 1$ is higher than the highest rank where $A = 0$, A is said to be *highly probable.*

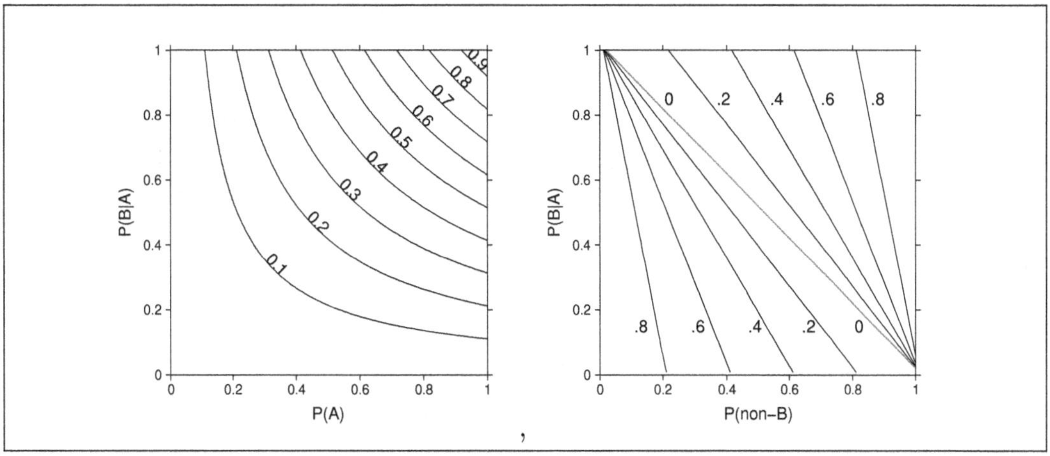

Figure 7: Contour lines of lower probabilities. Left: MODUS PONENS (equivalent to a product copula); the contour lines are increasing in the X- and in the Y-axis. Right: MODUS TOLLENS; points on the diagonal from (0,1) to (1, 0) correspond to $P(\neg A) = \theta' = 0$, above and to the right of the diagonal the lines are increasing in the X- and increasing in the Y-axis; below the diagonal they are decreasing in the X- and increasing in the Y-axis.

A conditional event $B|A$ is *highly probable* if the highest rank of those constituents where $(A \wedge B) = 1$ is higher than the highest rank of those constituents where $(A \wedge \neg B) = 0$. To test for *high probability* requires to first compute the probabilities of the constituents which generate the lower bound of the probability of the conclusion.

Definition 5 (High probability-preservation)**.** An inference rule is high probability preserving if all premises and its conclusion have high probabilities [4, 8, 9].

The uncertainty-sum condition and high probability preservation are equivalent [8, p.9, p.14ff.].

Figure 8 shows two examples, the CONJUNCTION and the MODUS TOLLENS with conditional event interpretation of the conditional. For all values of the probabilities of the premises from 0 to 1 in steps of .005 first the lower bound of the conclusion was found by linear programming. The solution of the linear program provided the probabilities of the constituents. The probabilities of the constituents were rank ordered and the premises and the conclusion were checked for *high probability*. The MODUS PONENS generates the same area as the MODUS TOLLENS on the right hand side of Figure 8.

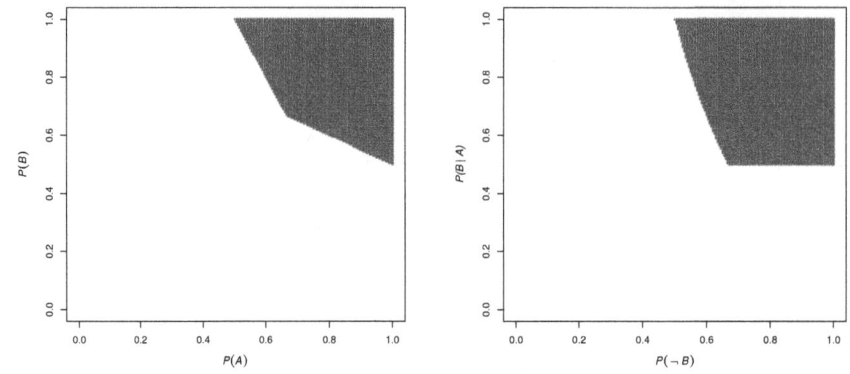

Figure 8: High probability-preservation. Left: CONJUNCTION; the dark area marks the cases where high probabilities of A and B lead to a high probability of the conclusion $P(A \wedge B)$. Right: MODUS TOLLENS with conditional event interpretation of the conditional; the dark area marks the cases where high probabilities of the premises, $\neg B$ and $B|A$, lead to a high probability of the conclusion $\neg A$.

High probability preservation is important for practical reasons [82, p.86]. Although "high probability" is not defined by a fixed numerical threshold, for every-day inferences the fixed value .5 will do a good job (compare [16]). For every-day inferences it is an intuitively appealing property. I am not aware, however, of an empirical study on high probability-preservation in human reasoning research. The criterion is defined in terms of the coherent probabilities of the constituents. Adams [8, p.14] remarks that '... the author [Adams] and others have taken this condition as 'the' criterion of validity for inferences with less-than-certain premises ...".

6 The interpretation of conditionals

Psychologists observed that humans often interpret conditionals as conditional events and not as material implications. The human understanding of conditionals seems to agree with Adams' conditional probability interpretation. Does human reasoning also agree with Adams' conception of p-validity? For Adams p-validity is a surrogate for classical validity. Is human uncertain reasoning sensitive to p-validity so that the classical distinction between valid and nonvalid rules can be transferred to uncertain reasoning? To better understand the intimate relationship between the interpretation of conditionals and p-validity we ask why the probability of the conjunction or a special form of the conjunction, called quasi-conjunction, of the premises can become a criterion for the validity of a probabilistic inference rule.

Adams [6] distinguishes formulas with and without conditionals. Conditional-free formulas are called *factual.* Let $X_1^f, \ldots, X_n^f$ and Y^f be such factual formulas. Probabilities in inference forms containing only factual formulas show a parallel to classical logic. In classical logic the *conjunction* of the premises of a valid argument implies its consequence:

$$\text{If } \{X_1^f, \ldots, X_n^f\} \models Y^f, \text{ then } \bigwedge_{i=1}^{n} X_i^f \rightarrow Y^f, \tag{15}$$

where $\models$ denotes entailment and $\rightarrow$ denotes material implication. The uncertainty-sum criterion is equal to the lower bound of the probability of the *conjunction* of the factual premises:

$$\gamma \geq P(X_1^f \wedge \ldots \wedge X_n^f) = \max\{0, \sum_{i=1}^{n} P(X_i^f) - (n-1)\}.$$

For factual languages (in which conditionals are expressed by material implications) Suppes [87] was the first who showed that inference forms that are valid in classical logic satisfy the uncertainty-sum criterion in probability logic. Adams extended the uncertainty-sum criterion to languages in which conditionals are expressed by conditional events.

Conditional events are not propositions and cannot be connected by the usual conjunction. To circumvent this restriction Adams [6, p. 46] [9, p.164ff.] introduced as an *Ersatz* the *quasi-conjunction.* It is defined by

$$C(B_1|A_1, B_2|A_2) = (A_1 \rightarrow B_1) \wedge (A_2 \rightarrow B_2) \mid (A_1 \vee A_2), \tag{16}$$

where A_1, B_1 and A_2, B_2 are unconditional events and A_1 and $A_2 \neq \bot$. The definition involves two kinds of conditionals, (a) two material implications, $(A_1 \rightarrow B_1)$ and $(A_2 \rightarrow B2)$, and (b) one conditional event signaled by the vertical stroke. Adams [6, p. 46] calls the right arrows (he uses the $\supset$ notation) the "material counterparts" of the conditional events in a set of premises or conclusions. Now assuming the disjunction $A_1 \vee A_2$ to be true, we can build the conjunction of two material implications "as usual". Adams uses this construction for the *entailment* relation, $\models$, in probabilistic inference forms involving conditional events. Let $P(B_1|A_1)$ and $P(B_2|A_2)$ be two premises and assume $A_1 \vee A_2$ to be true. The probability of the quasi-conjunction of the premises $P(C(B_1|A_1), B_2|A_2))$ becomes the criterion for the validity of an inference. Thus, the conditional event interpretation refers to the *entailment relation,* not to the conditionals "inside" the premises! This led to serious misunderstandings in the psychological literature.

An inference rule is p-valid if the probability of its conclusion is at least as probable as the probability of the quasi-conjunction of its premises. To obtain the lower bound of the quasi-conjunction (i) the conditionals in the premises are replaced by their material counterparts and (ii) the lower bound of the probability of their conjunction is determined under the constraint that the disjunction of the conditioning events in the premises is true. We are familiar with the result: The lower bound is just the lower bound of the conjunction of the premises when the conditional events are replaced by material implications and we have seen that this is the uncertainty-sum criterion of p-validity.

Consider a probabilistic inference rule R in a language containing conditional events with the form

$$P(X_1) \in [\alpha'_1, 1], \ldots, P(X_n) \in [\alpha'_n, 1] \models P(Y) \in [\theta', 1], \ 1 \leq n < \infty.$$

The premises $X_1, \ldots, X_n$ and the conclusion Y contain N elementary events $A_1, B_1, \ldots, A_N, B_N$. Assume the premises are logically independent so that they generate 2^N constituents. We encode the the premises in an $(n+1) \times 2^N$ matrix $\mathbf{Q}$. We follow Adams and replace the conditional events $B_1|A_1, \ldots, B_n|A_n$ by their material counterparts $A_1 \rightarrow B_1, \ldots, A_n \rightarrow B_n$ as prescribed by the definition in (16). The first n rows of $\mathbf{Q}$ contain the 0/1 truth values of the premises. In row $n+1$ we put the truth values of the disjunction $A_1 \vee \ldots \vee A_n$. The disjunction functions as the sure event Ω with $P(A_1 \vee \ldots \vee A_n) = 1$. In addition we build the column vector $\mathbf{b} = (\alpha'_1, \ldots, \alpha'_n, 1)^T$ containing the probabilities of the premises and the disjunction probability 1. The conclusion Y is encoded in the objective function of a linear program. If for all coherent probability assessments of the premises (including the constraint $P(A_1 \vee \ldots \vee A_n) = 1$) the linear system built by $\mathbf{Q}$, $\mathbf{b}$, and the objective function defined by the conclusion Y has a solution with $\gamma' \geq \max\{0, \sum_{i=1}^{n} \alpha'_i - (n-1)\}$ for $P(Y)$, then the premises entail the conclusion and the inference rule is p-valid.

To see this let $K = \{j : \sum_{i=1}^{(n+1)} q_{ij} = n+1\}$ be the set of all columns of $\mathbf{Q}$ which contain 1s only. This set represents the quasi-conjunction of the premises. Assume $K \neq \emptyset$.[6] Next consider the truth values y_j, $j = 1, \ldots, 2^N$, of the conclusion Y. In a logically valid inference form the conclusion vector must not contain a 0 in position j and $j \in K$. As $P(Y) = \sum_{j=1}^{2^N} y_j x_j$, where x_j denotes the unknown probability of the $j^t h$ constituent, $P(Y)$ must at least be equal to $P(K)$. Thus the probability of the quasi-conjunction K is the lower probability that the premises entail the conclusion.

Gilio [37, 41, 42] used a tri-valued valuation of the premises. This leads to the same lower bound of the quasi-conjunction as above. Goodman, Nguyen, and Walker

[6] The premises are consistent. We come back to this property in section 7.

[44, 61] investigated operators combining conditional events in conditional event algebras, structures that are extensions of Boolean algebras. Quasi-conjunction was investigated in a possibilistic setting by Benferhat, Dubois and Prade [15]. For a criticism of the quasi-conjunction see [22]. The quasi-conjunction is a rather permissive bound that may easily lead to incoherent results if used to evaluate individual probability judgments. For a probabilistic semantic of a default logic Coletti and Scozzafava [22, chapter 20] proposed to work with probability 0 and 1 assessments only. They showed that the proposal sanctions all rules of System P and rejects monotonicity, contraposition, and transitivity. In several of our papers [69, 70] we considered inferences with a *noninformative* probability interval $[0, 1]$ of the conclusion as inconclusive. Monotonicity, contraposition, and transitivity are probabilistically noninformative.

P-validity represents a semantics for System P. This has been shown for propositional languages, e.g., in [9, 84] and for languages extended by conditionals in [6]. System P has three purely probabilistic axioms: *cautious monotonicity*

$$\text{from } \{P(B|A), P(C|A)\} \text{ infer } P(C|A \wedge B),$$

cautious cut or *cautious transitivity*

$$\text{from } \{P(B|A), P(C|A \wedge B\} \text{ infer } P(C|A),$$

and *cautious disjunction*

$$\text{from } \{P(C|A), P(C|B)\} \text{ infer } P(C|A \vee B).$$

Consider *cautious monotonicity.* If we drop A, which is a conditioning event in the two premises and in the conclusion, then *cautious monotonicity* is equivalent to *conditioning*: From $\{P(B) = \alpha_1,\ P(C) = \alpha_2\}$ infer $P(C|B) = \gamma \in [\frac{\alpha_1+\alpha_2-1}{\alpha_1}, \frac{\alpha_2}{\alpha_1}]$. Moreover, after dropping A *cautious transitivity* is equivalent to the *modus ponens.*

The disjunction rule is best known from Simpson paradox [65]. Take as an especially easy example $P(C|A) = P(C|B) = \alpha$; then the lower bound of $P(C|A \vee B) = \gamma' = \alpha/(2 - \alpha)$. If $\alpha = .6$, then $z' = .4286$. So it may happen that a therapy helps 60 % of male and 60 % of female patients but overall it helps only 45 % of all patients. The uncertainty-sum criterion is $.6 + .6 - 1 = .2$, i.e., much below the coherence bound.[7] While many psychological experiments investigated the "modus-ponens-quartet", only a few studies investigated the axioms of System P (see however [69, 73, 72]). The *modus ponens* is part of both settings.

System P is sound and complete. All that is required to decide the p-validity of an inference are the probabilities of the premises, no intermediate "derivations" are needed; "...we may compute the lower bound of the entire inference in *one*

[7]The Simpson paradox is "resolved" if the lower and upper probabilities are expressed as second order distributions and the probability of being coherent is maximized [49]. A simple approximation is the midpoint of the $[\gamma', \gamma'']$-interval, in the example $(\gamma' + \gamma'')/2 = .589$. That is the probability for all patients is approximately the same as for the male and for the female patients.

step, independently from the way in which the conclusion has been proved from the premises." [81, p.542]

In a psychological study we should state what we want the participants to do: (i) to evaluate the validity of a rule or (ii) to evaluate the probability of a conclusion. The computation of an uncertainty-sum is in fact "easy". To check the p-validity of a rule, however, the uncertainty-sum criterion (γ' in Definition 2) must be compared with all (!) coherent lower probabilities (θ' in Definition 2) to decide whether $\gamma' \leq \theta'$ or not. Computationally this is expensive and psychologically unrealistic. It is however possible that such rules were reinforced, implemented, and hard-wired in our brain by evolution [83].

Most probability assessments in every-day life are made in terms of point probabilities and not in terms of interval probabilities with upper bounds equal to 1. Similarly, in reasoning research experimenters ask participants to provide point probability judgments of the premises and not interval probabilities with upper bounds 1. For the comparison of human judgments with the formal model the interval probabilities of the premises should be replaced by point probabilities. The probability of the conclusions are still interval probabilities. The lower bounds are the same as in the previous sections, but the upper bounds are now usually less than 1. As a consequence, experiments asking for the assessment of point probabilities of the premises, like [86, 31], should evaluate the judgments of their participants with the correct upper bounds and not with upper bounds equal to 1.

The lower and upper bounds of interval probabilities are related by the *conjugacy* property (in the context of imprecise probabilities see [90, 12, 78], in the context of the coherence approach see [21]):

> *The upper bound of an interval probability of an event A is equivalent to 1 minus the lower bound of its complement, $P''(A) = 1 - P'(\neg A)$.*

For a precise probability assessment of the premises the p-validity criterion becomes:

Definition 6 (P-validity for precise probabilities). *Let R be a probabilistic inference rule with the form*

$$P(X_1) = \alpha_1, \ldots, P(X_n) = \alpha_n \quad \models \quad P(Y) \in [\gamma', \gamma''],$$

where the X_i are logically independent. R is p-valid for point probabilities if the degree of essentialness of each of the premises is $e_i = 1$, $i = 1, \ldots, n$, and if the probability of the conclusion $\gamma = P(B)$ is in the interval

$$\gamma \in \left[\max\left\{0, \sum_{i=1}^{n} \alpha_i - (n-1)\right\}, \min\{\alpha_1, \ldots, \alpha_n\} \right]. \tag{17}$$

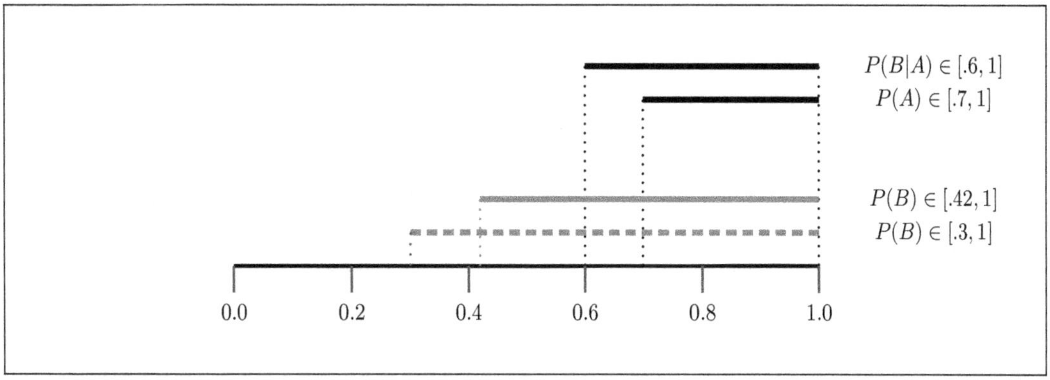

Figure 9: MODUS PONENS with the interval premises $P(A) \in [.7, 1]$ and $P(B|A) \in [.6, 1]$. The coherent probability of the conclusion is $P(B) \in [.42, 1]$. The uncertainty-sum criterion leads to $P(B) \in [.3, 1]$. Probabilities in the interval $[.3, .42]$ are incoherent.

If $e_i \neq 1$, $i = 1, \ldots, n$, *the lower bound is given by Equation (6) and the upper bound is obtained by the conjugacy property.*

That is, γ is within the bounds of the conjunction of a set of premises and the premises are either conditional-free or the conditionals in the premises are interpreted as material implications. The lower and upper bounds in (17) are the Frechét-Hoeffding copulas. They hold when no assumptions about the dependence or independence of the events are made. The bounds were already known to George Boole [20, p.298/9] who gave credit to de Morgan.

Figure 9 shows an example of a MODUS PONENS in which the premises were assessed by interval probabilities. Figure 10 shows a closely related example with a point probability assessment. On the left side of Figure 9 there is a region of incoherence resulting from the difference of g-coherence and coherence. In Figure 10 there are two regions of incoherence, one on the left and one on the right side. They result from applying the uncertainty-sum criterion to point probabilities of the premises. If in an experiment the participants assess point probabilities of the premises and the responses are evaluated by the uncertainty-sum criterion, then incoherent responses may be classified as "rational".

Because of the conjugacy property p-validity for point probabilities may equivalently be defined in terms of upper probabilities: R is p-valid if $P(B) \leq \gamma''$. Here p-validity not only protects against too low but also against too high probability judgments. Evaluating human judgments using upper probability 1 when the premises have point probabilities ignores the gap between γ'' and 1.

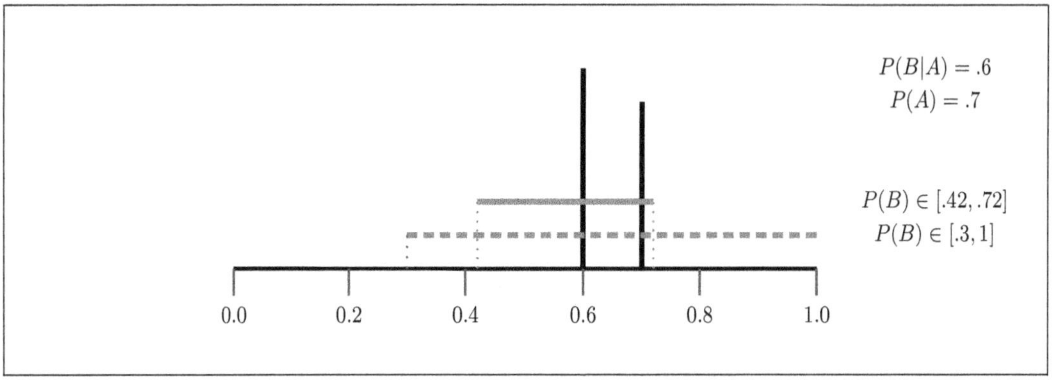

Figure 10: MODUS PONENS with the point premises $P(A) = .7$ and $P(B|A) = .6$. The coherent probability of the conclusion is $P(B) \in [.42, .72]$. The uncertainty-sum criterion leads to $P(B) \in [.3, 1]$. Probabilities in the intervals $[.3, .42]$ and $[.72, 1]$ are incoherent

A different but closely related question arises: under which conditions can upper probabilities equal to 1 be coherent at all. We turn to this question in the next section.

7 Inconsistent premises

The upper probabilities of the premises of a p-valid inference form are all equal to 1, $\alpha \in [\alpha_i, 1], i = 1, \ldots, n$. This can only be coherent if at least one column in the matrix of coefficients $\mathbf{Q}$ contains 1s only. If all entries in column j are 1 then the conjunction of the premises is true if $x_j = 1$, i.e., $x_j = 1$ leads to the existence of a solution of the linear system. If $\boldsymbol{Q}$ does not contain such a column then the premises are *inconsistent*. In this case an assessment with all probabilities equal 1 is incoherent. Thus p-validity requires consistent premises.

Here an example of inconsistent premises given by Levine and Adams [11, p. 435].

Example 7 (Inconsistent premises). *From* $P(A) = \alpha_1, P(B) = \alpha_2, P[\neg(A \wedge B)] = \alpha_3$ *infer* $P(A \leftrightarrow \neg B) = \gamma$.

Table 4 shows the possible indicator values of the premises and the conclusion. The premises are inconsistent as there is no column containing only 1s. The proba-

	x_1	x_2	x_3	x_4	α_i
A	1	1	0	0	α_1
B	1	0	1	0	α_2
$\neg(A \wedge B)$	0	1	1	1	α_3
Ω	1	1	1	1	
$A \leftrightarrow \neg B$	0	1	1	0	γ

Table 4: Inference with inconsistent premises (Adams [11, p.435]).

bility of the conclusion is obtained by solving

$$\begin{aligned} x_1 + x_2 &= \alpha_1 \\ x_1 + x_3 &= \alpha_2 \\ x_2 + x_3 + x_4 &= \alpha_3 \quad \textit{and} \quad \sum_{i=1}^{4} x_i = 1, x_i \geq 0, i = 1, \ldots, 4. \end{aligned}$$

for the objective function $\gamma = x_2 + x_3$. *The solution is precise:*

$$\gamma = \alpha_1 + \alpha_2 - 2(1 - \alpha_3)\,,$$

where $\alpha_1 + \alpha_2 \geq 2(1 - \alpha_3)$ *is required for coherent premises. Upper probabilities 1 would be incoherent.*

The MODUS TOLLENS is another example where the premises are inconsistent.

To find lower and upper probabilities Adams and Levine introduce several subsets of the premises, such as "minimal sufficient sets", "minimal essential sets", and "negatively sufficient sets". With the help of these sets they build a "minimal falsification matrix" which is solved by linear programming. A critical discussion of Levine and Adams is given in Hailperin [46, p.168 ff.]. He "...show[s] how Adams-Levine's minimal falsification matrix ... can be obtained by a straightforward (to us, less obscure) method, Boole's 'purely algebraic form' ..." [46, p. 171].

If the premises contain conditional events having probability 1, then the entries in the matrix of coefficients are the same as if the conditional is interpreted as a material implication. We remember that for conditional events with false antecedents the entries are conditional probabilities which are now equal to 1. This corresponds exactly to the truth function of the material implication; its valuation is TRUE if the antecedent is FALSE.

P-validity may be seen as a relation between two interpretations of conditionals, conditional event and material implication. Here is a close link to the new probabilistic paradigm in reasoning research in which the human interpretation of conditionals is a key topic. Strong evidence was found that conditionals are interpreted as conditional events [32, 50]. We compare the two interpretations for the INTRODUCTION OF A CONDITIONAL, that is, for the inference from $\{A, B\}$ to "if A then B".

Example 8 (INTRODUCTION OF MATERIAL IMPLICATION). *If* $P(A) = \alpha_A$ *and* $P(B) = \alpha_B$, *then the probability of the material implication* $P(A \to B) = \gamma_\to$ *is in the interval*

$$\gamma_\to \in [\max\{1 - \alpha_A, \alpha_B\}, \min\{1, 1 - \alpha_A + \alpha_B\}]. \tag{18}$$

Because $P(A \to B) = 1 - P(A \wedge \neg B)$ *the minimum of* $\gamma_\to$ *is obtained if the probability of the conjunction of* A *and* $\neg B$ *is maximal,* $P(A \wedge \neg B) = \min\{\alpha_A, 1 - \alpha_B\}$ *so that its 1-complement is* $\gamma'_\to = \max\{1 - \alpha_A, \alpha_B\}$. *The maximum is obtained if* $P(A \wedge \neg B)$ *is minimized, i.e., if* $P(A \wedge \neg B) = \max\{0, \alpha_A + (1 - \alpha_B) - 1\}$ *so that the 1-complement is* $\gamma''_\to = \min\{1, 1 - \alpha_A + \alpha_B\}$.

Example 9 (INTRODUCTION OF A CONDITIONAL EVENT). *If* $P(A) = \alpha_A$ *and* $P(B) = \alpha_B$, *then the probability of the conditional event* $P(B|A) = \theta_|$ *is in the interval*

$$\theta_| \in \left[\max\left\{0, \frac{\alpha_A + \alpha_B - 1}{\alpha_A}\right\}, \min\left\{1, \frac{\alpha_B}{\alpha_A}\right\}\right], \quad \textit{if} \quad \alpha_A > 0 \tag{19}$$

$$\textit{and } \theta_| \in [0, 1] \textit{ if } \alpha_A = 0.$$

For $0 < \alpha_A \leq 1$ *the interval is obtained from* $P(B|A) = P(A \wedge B)/P(A)$ *and the bounds of the conjunction* $P(A \wedge B) \in [\max\{0, \alpha_A + \alpha_B - 1\}, \min\{\alpha_A, \alpha_B\}]$.

For probabilities close to .5 the intervals for the two interpretations can be remarkably different. Assuming $\alpha_A = \alpha_B = .5$, for example, leads to $\gamma_\to \in [.5, 1]$ for the material implication and to $\theta_| \in [0, 1]$ for the conditional event interpretation. To infer from $\alpha_A = .5$ and $\alpha_B = .5$ that the probability of "if A then B" is greater than .5 seems to be absurd; to infer the vacuous interval $[0, 1]$ seems to be reasonable. I am not aware of an empirical study comparing the two interpretions along this line. A related question about the INTRODUCTION OF CONDITIONAL EVENTS leads to the next section.

8 The null-unity convention

In the psychological literature the probability of premises are sometimes taken to be equal to one [62, 63]. In the probabilistic version of DENYING THE ANTECEDENT, "from $\{P(\neg A), P(B|A)\}$" infer "$P(\neg B) = \ldots$", it may be assumed that $P(\neg A) = 1$. This section deals with questions that arise when the probabilities of the conditioning events are 0 or 1, i.e., questions about $P(B|A)$ when $P(A) = 0$ or $P(\neg A) = 1$, respectively.

Kolmogorov [51] formulated probability axioms for *elementary events* like A and B. He introduced conditional probability by the ratio definition

$$P(B|A) = \frac{P(A \wedge B)}{P(A)}, \quad \text{for } P(A) > 0\,.$$

In a way statements about conditional statements are abbreviations for longer statements about unconditional probabilities [14, p.18]. If $P(A) = 0$ the ratio is undefined. As a consequence, if $P(A) = 1$, then $P(B|\neg A)$ is also not defined. This means that in such cases inference rules like the MODUS PONENS or the MODUS TOLLENS are undefined, in a psychological context a painful restriction.

The coherence approach (like several other approaches, best known perhaps the approach of Popper [76, 77]) starts from axioms for *conditional events* like $B|A$ [22, p.73/74]. Conditioning on events with probability zero is completely legal [22]. A conditional probability is an element in the set of all values t that satisfy the axiom of compound probability

$$P(B|A) \in \{t \in [0,1] : P(A \wedge B) = tP(A)\}$$

(compare [14, p. 18]). For $P(A) = 0$ this consists of the set of all real numbers in $[0,1]$, i.e., the "vacuous" interval. The coherence approach, however, does not allow conditioning on the impossible event $\emptyset$, i.e., $P(A|\emptyset)$ is not allowed [22, p.63]. (Conditioning on the contradiction is legal in Popper's approach, where $P(B|A \wedge \neg A) = 1$ [77, p. 273].)

For $P(A) = 0$ Adams [4, 9] set $P(B|A)$ equal to 1. This has been called the "null-unity convention" [14, p. 18]. It takes only the maximum of the $[0,1]$-interval and thus replaces maximal imprecision with the highest precise value, substituting certainty for the vacuous interval. As a consequence, for $P(A) = 0$ we have $P(B|A) + P(\neg B|A) = 1 + 1 = 2$ which is absurd. (A similar problem appears in Popper's approach where $P(A|A \wedge \neg A) + P(\neg A|A \wedge \neg A) = 2$ [77, p.305]).

The motivation behind the null-unity convention is to export the role of the material implication in the consequence relation from deductive logic to probability logic. The truth values FALSE and TRUE lure behind probabilities 0 and 1. The truth value of the material implication $A \to B$ is a function of the truth values of A and of B. $A \to B$ is TRUE if A and B are TRUE and if A is FALSE. It is FALSE if A is TRUE and B is FALSE. What "if A is false, then $A \to B$ is true" is in logic becomes "if $P(A) = 0$, then $P(B|A) = 1$" in probability logic.

The probability of a conditional event $B|A$ is constrained by the probability of A and B such that if $P(A) = x, x > 0$ and $P(B) = y$, then $P(B|A) = z$ is in the interval

$$z \in [\max\{0, \frac{x+y-1}{x}\}, \min\{\frac{y}{x}, 1\}] \,. \tag{20}$$

This is illustrated in Figure 11. It shows the lower and upper probabilities for four selected values $P(B) = .2, .4, .6$ and $.8$. $P(A)$ is represented on the X-axes. The lower and upper bounds of $P(B|A)$ are shown by the dotted and the solid lines, respectively. For a fixed value of $P(A)$ all $P(B|A)$ in the interval between the dotted and the solid lines are coherent. As $P(A)$ approaches zero the lower and upper probabilities approach the unit interval $[0, 1]$. The small circle at $P(A) = 0$ and $P(B|A) = 1$ indicates the null-unity convention. It replaces "probabilistic noninformativeness" by "certainty". The null-unity proposal establishes an analogy between the truth values of a material implication and probability 0 and 1 of a conditional event. This may also be seen from the fact that if $P(A) = 0$ fixes the value of $P(B|A)$ at 1, this amounts to a perfect analogy to "from $\neg A$ infer $A \to B$", the paradox of the material implication. Since the probability of the premise is $P(\neg A) = 1$ and the probability of the conclusion is $P(B|A) = 1$, the paradox would be p-valid for this special case.

The lower and upper probabilities in Figure 11 show how misleading this analogy is. In the realm of probability a decreasing $P(A)$ leads to increasing *ignorance* and clearly not to certainty.

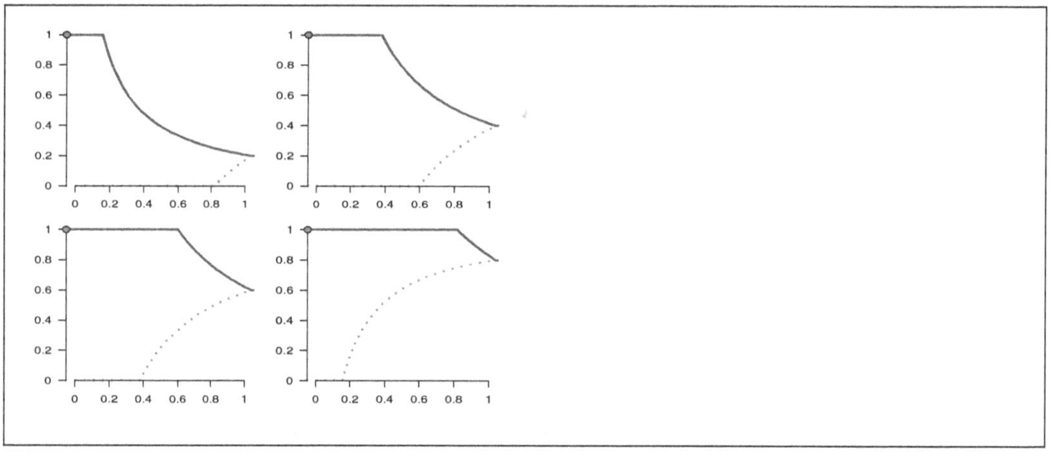

Figure 11: Lower (dotted) and upper (solid) conditional probabilities $P(B|A)$ (on the Y-axis) as functions of $P(A)$ (on the X-axis). Four different values of $P(B) = .2$ (top), $.4, .6$ and $.8$ (bottom). Setting $P(B|A) = 1$ if $P(A) = 0$ projects the $[0, 1]$ intervals to just one point (small circle at $P(B|A) = 1$).

Zero probabilities have an impact on the quartet of the categorical syllogisms (see Table 1). If in a MODUS PONENS the unconditional premise is $P(A) = 0$ then the ratio definition of conditional probabilities makes the second premise $P(B|A)$ undefined. No conclusion can be drawn. If we take the null-unity convention, then $P(B|A)$ is fixed at $\beta = 1$, i.e., the probability of the conditional premise is not free to vary. The lower and upper bounds, $\alpha\beta$ and $1 - \alpha + \alpha\beta$ of $P(B)$ are 0 and 1, i.e., the inference is probabilistically noninformative. In the coherence approach the inference is also probabilistically noninformative, but $P(B|A)$ is free to vary. If $P(A) = 0$. The probabilistic MODUS PONENS is probabilistically noninformative for *all* $P(B|A) \in [0, 1]$.

In the MODUS TOLLENS the upper bound of the conclusion $P(\neg A)$ is $\gamma'' = 1$ (see Table 1). $P(\neg A) = 1$ implies $P(A) = 0$ and with the ratio definition the conditional premise $P(B|A)$ is not defined. If we apply the null-unity convention the value $\gamma'' = 1$ implies that $P(B|A) = \beta = 1$ (since $P(A) = 0$) and is not free to vary. In the coherence approach $\gamma'' = 1$ is the upper bound for all $\beta \in [0, 1]$. Similar results are obtained for DENYING THE ANTECEDENT and AFFIRMING THE CONSEQUENT.

In the coherence approach it is completely legal to work with the zero probability of the conditioning event, i.e., with $P(B|A)$ where $P(A) = 0$. In the INTRODUCTION OF CONDITIONAL EVENTS it leads to the vacuous interval $\gamma_| \in [0, 1]$. In more complex cases the computational investigation of zero probabilities involves two steps:

1. Find out whether the zero probability of the conditioning event is compatible with the linear system. To do so, introduce the conditioning event as a new consequent and find its lower probability. In a linear program, e.g., the event is represented by a new objective function. If the lower bound of the consequent is zero, the zero probability of the conditioning event is compatible.

2. If the zero probability is compatible, include the conditioning event along with its zero probability explicitly in the system and solve it for its original consequent or objective function.

The steps may be iterated leading to a sequence of "zero layers" [22] (for alternative algorithms see [92] or [23]).

The psychological literature on zero probabilities is confusing. Oaksford and Chater [62, p. 111], e.g., consider the following paradox of the material implication,

$$\text{from } \neg A \text{ infer } A \to B.$$

It is counter-intuitive, but valid in classical logic. Its probabilistic versions reads

$$\text{from } P(\neg A) = \alpha \text{ infer } P(B|A) = \beta.$$

First, Oaksford and Chater cite Bennett [17]: "your conditional probability for B given A is the probability for B that results from adding $P(A) = 1$ to your belief system ..." [changed notation][8],[9]. They continue to follow Bennett and speak of a "zero-intolerance" if $P(A) = 0$. They conclude that "... when $P(A) = 1$, then $P(A \to B)$ should also be 1."

Next the authors argue that if $\alpha = 0$ the ratio formula is undefined[10] and "... no value can be assigned to the probability of the conditional in the conclusion because of *zero-intolerance*." [62, p. 111]. If $\alpha = 1$, then β should also be 1, "... which means their uncertainties are equal. However, p-validity requires the inequality in 5.8 [an equation on the same page] to hold whatever the value of α [changed notation]." The formula 5.8 follows Adams [9, p. 187]), however, and defines p-validity in terms of "less than or equal" ($\leq$) and not in terms of "less than" ($<$). Moreover, Adams suppresses the paradox not because of zero probability, but because it may happen that the "... inference has a highly probable premise and a highly improbable conclusion, which would make it seem irrational for persons to reason in this way, at least if they hope to reach probable conclusions thereby." The probability of the conclusion is not a monotonically increasing function of the probability of the premise (not 1-increasing in the sense of section 5).

In the coherence approach the suppression of the paradoxes of the implication question is straightforward: If $P(A) = 0$, then $P(B|A) \in [0, 1]$, that is, the inference is *probabilistically noninformative.* If you think that the probability of A is zero, you may bet any amount because you think that you will always get your money back with probability 1.

In the four inference forms MODUS PONENS etc. the probability of the conditioning event is already stated in the premises. Its probability is not constrained by the second premise, so the lower value zero is compatible [33, p. 167] and $P(B)$ may be obtained from $P(B) = P(A)P(B|A) + P(\neg A)P(B|\neg A)$. The first term vanishes and as $P(B|\neg A)$ is not constrained $P(B) \in [0, 1]$. We note that "... by defining conditional probability as any solution to equation $P(A \cap C) = P(A|C) \cdot P(C)$, it still makes sense as a non-negative number when $P(C) = 0$ (see details in Coletti and Scozzafava ...)" [28, p.15]

The investigation of the "conditioning on zero probability events" was often motivated by paradoxes such as the Borel-Kolmogorov paradox [45] or the first digit problem [22]. An important concept is the conglomerability property, first described by de Finetti [25]. Important contributions were provided by [27, 80, 100]. For the

[8]Page 109 in [62] contains a number of more or less obvious misprints.

[9]Ramsey did not propose to add the probability $P(A) = 1$ to the belief system but the proposition A.

[10]For $P(A) = 0$ Adams, however, takes $P(B|A) = 1$ and not as undefined!

treatment of zero probabilities in the theory of imprecise probabilities see [90, p. 306ff., p. 328ff.] and the references in [12, p. 47]. Adams was well aware of the zero-antecedent problem. He remarked that "... the antecedents of the conditionals involved may have zero probability and we have no theory which applied to that case." and he continued in a note "... the desirability of serious investigation of the zero-antecedent case ..." [6, p. 40 and note 5, p.41].

Singmann et al. [86] and Evans et al. [31] employed the lower bound $\max\{0, \alpha + \beta_? - 1\}$ to evaluate the probability judgments of the participants in their experiments. We use the question mark in $\beta_?$ to indicate that the interpretation of the conditional is unclear: "For generality and to minimize our assumptions we did not presuppose that $P(\text{if } p \text{ then } q) = P(q|p)$ in our assessment of p-validity." [31]. Not specifying the interpretation of conditionals however degrades the uncertainty-sum criterion to an adhockery. The lower bounds are coherent lower probabilities for the conjunction of the premises where conditionals are material implications, i.e., $\beta_?$ is—if coherent—actually $\beta_\rightarrow$. If in a psychological investigation p-validity intervals are determined, it would be consistent to determine the upper bounds with the material implication interpretation, that is, to work explicitly with $\gamma''_\rightarrow = 1$ and not with $\gamma''_? = 1$.

9 Correlated events

In a psychological context the judgment of correlations is often of similar importance as the judgment of probabilities. In an inference rule correlations may appear at two locations: at the premises and at the conclusion. We first turn to inferences about correlations, i.e., to correlations at the conclusion.

We use the notation of Table 2 and consider the 2×2 correlation ρ between the

	B	$\neg B$	$\sum$
A	x_1	x_2	$\alpha_A = x_1 + x_2$
$\neg A$	x_3	x_4	$1 - \alpha_A = x_3 + x_4$
$\sum$	$\alpha_B = x_1 + x_3$	$1 - \alpha_B = x_2 + x_4$	1

Table 5: Notation in the 2×2 scheme to calculate ρ

two binary events A and B,

$$\rho = \frac{x_1 x_4 - x_2 x_3}{\sqrt{\alpha_A(1 - \alpha_A)\alpha_B(1 - \alpha_B)}} \,. \tag{21}$$

The marginal probabilities α_A and α_B constrain the value of ρ; lower and upper bounds of ρ are obtained by (21) with the help of the conjunction probabilities

$x_1 \in [\max\{0, \alpha_A + \alpha_B - 1\}, \min\{\alpha_A, \alpha_B\}]$, $x_2 = \alpha_A - x_1$, $x_3 = \alpha_B - x_1$, and $x_4 = 1 - (x_1 + x_2 + x_3)$.

What is the relationship between conditionals and correlation in probabilistic inference? In the same way as we may infer from a set of premises the lower and upper probabilities of conclusions, we may infer lower and upper correlations. Are there systematic differences between correlations inferred from p-valid and inferred from p-nonvalid schemata? What a difference makes the interpretation of conditionals—conditional event or material implication—on correlational inferences?

The probability of a single *if A then B* sentence that is interpreted as a conditional event carries no information about the correlation between the two events. From $P(B|A) = \beta_|$ we can only infer the vacuous interval $\rho_| \in [-1, 1]$. From the material implication $P(A \to B) = \beta_\to$ we infer $\rho_\to \in [-1, \frac{\beta_\to/2}{1-\beta_\to/2}]$. The lower bound is obtained if $x_1 = x_4 = 0$ in Table 5. The upper bound is obtained if $x_3 = 0$ and if the numerator in (21) is maximized. This is the case if $x_1 = x_4 = \beta_|/2$ so that the product $x_1 x_4$ in the numerator of (21) obtains a maximum. If, for example, $\beta_\to = 0$ then $\rho_\to \in [-1, 0]$, if $\beta_\to = .5$ then $\rho_\to \in [-1, 1/3]$, and if $\beta_\to$ is close to 1 the interval of the correlation becomes vacuous.

We next consider inferences about ρ from the premises of the MODUS PONENS etc., both for the material implication interpretation, $[\rho'_\to, \rho''_\to]$, and for the conditional event interpretation, $[\rho'_|, \rho''_|]$, of the conditional in the premises. For the 2×2 case results are obtained with the help of the lower and upper probabilities of the conclusions, i.e., $P(B), P(\neg A), P(\neg B)$, and $P(A)$. They allow to determine x_1, x_2, x_3, and x_4 which are required to determine ρ by Equation (21). Figure 12 shows two numerical examples, one for the MODUS PONENS and one for the MODUS TOLLENS. The probabilities of the minor premises, i.e., $P(A)$ and $P(\neg B)$, respectively, are fixed at $\alpha = 0.5$. Because of the conjugacy property the results for the MODUS PONENS and DENYING THE ANTECEDENT are identical and the same holds for MODUS TOLLENS and AFFIRMING THE CONSEQUENT.

For all four inferences the upper correlation increases from $\rho_\to = 0$ at $\beta_\to = .5$ approximately linearly up to $\rho_\to = 1$ at $\beta_\to = 1$. At $\beta_\to = .5$ the correlation can only be negative, at $\beta_\to = 1$ it can only be positive. As coherence requires $\beta_\to \geq \alpha$ and in the examples $\alpha = .5$ the $\rho_\to$ is undefined for $\beta_\to \leq .5$.

For the conditional event interpretation the MODUS PONENS and DENYING THE ANTECEDENT $\rho_|$ increases approximately linearly from $[-1, 0]$ up to $[0, 1]$ as $\beta_|$ increases from 0 to 1. At $\beta_| = \alpha$ the value of $\rho_|$ is undefined. For the MODUS TOLLENS and AFFIRMING THE CONSEQUENT lower and upper correlations switch symmetrically their values below $\beta_| = .5$ and above $\beta_| = .5$, respectively. At $\beta_| = \alpha = .5$ the lower and upper correlations coincide at $\rho_| = 0$.

With the premises of the MODUS PONENS the lower and upper bounds of the correlation for the conditional event interpretation are monotonically increasing from $[-1, 0]$ to $[0, 1]$. The bounds are symmetrical around $[-.5, .5]$ and the width of the interval is constant. ρ' and ρ'' are monotonically increasing with $P(B|A) = \beta_|$. These properties appear intuitive and reasonable. With the premises of the MODUS TOLLENS and $P(B|A) < .5$ the correlation must be negative; the lower bound is monotonically increasing and the upper bound is $\rho'' = 0$. With $P(B|A) > .5$ the relations flip symmetrically, the lower bound remains constant at $\rho' = 0$.

For the material implication interpretation the coherence of the premises requires that $P(A \to B) = \beta_\to \geq 1 - \alpha$. In the example with $\beta_\to = .5$ the results for the MODUS PONENS and the MODUS TOLLENS are very similar with monotonically increasing intervals from $[-1, 0]$ at $\beta_\to = .5$ up to $[0, 1]$ at $\beta_\to = 1$.

Inferences about lower and upper correlations allow to investigate qualitative properties like "the correlation is positive" or "the correlation is negative", respectively. In the probabilistic approach to reasoning such qualitative properties were elegantly investigated for the probabilities in inference rules [75]. We next turn to the case of correlated premises.

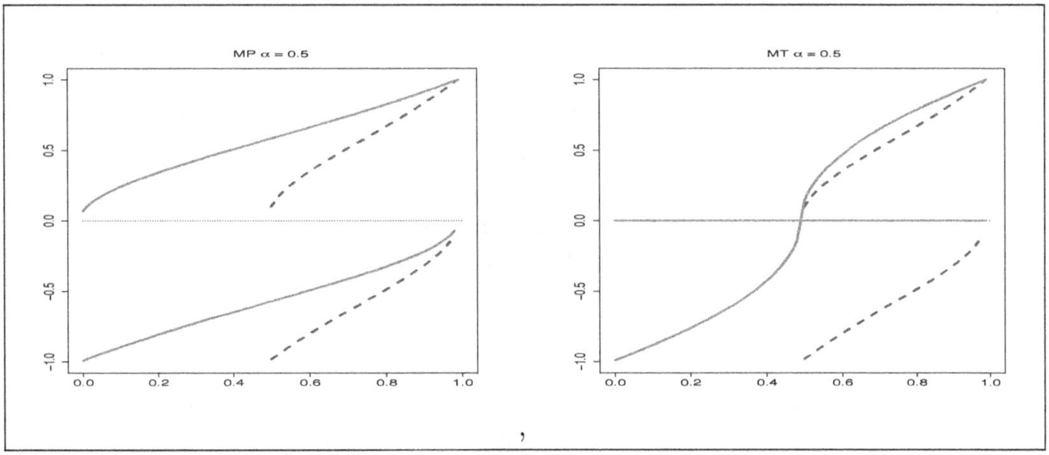

Figure 12: Lower and upper 2×2 correlation ρ on the Y-axis; $P(A)$ fixed at $\alpha = .5$; $\beta = P($ if A then $B)$ from 0 to 1 on the X-axis. Left: MODUS PONENS. Right: MODUS TOLLENS. Solid line: Conditional event interpretation, dashed: Material implication interpretation.

Experiments on human reasoning often investigate inferences with *content-lean* material, like "If there is an A on one side of the blackboard, then there is a B on the other side.". Such conditionals do not carry information about the dependence or independence of the involved events. If however content-rich material is presented, then the participants have background knowledge that will enter the inference process. Especially *if-then* sentences in the premises will activate beliefs about causal and correlational dependencies. Similarly, in every-day arguments events are usually supposed to be correlated.

If in a MODUS PONENS task the participants assess the point probabilities $P(A) = \alpha$ and $P(B|A) = \beta_|$ and *infer the point probability* $P(B) = \gamma_| = \beta_|$ and thus ignore the base rate $P(A)$, then they assume (implicitly or explicitly) that A and B are probabilistically independent. Likewise, for any other point probability of the conclusion it is easy to infer the corresponding correlation. For the MODUS PONENS with conditional event interpretation we apply Equation (21) with

$$\begin{aligned} x_1 &= \alpha\beta_| & x_2 &= \alpha(1-\beta_|) \\ x_3 &= \gamma_| - x_1 & x_4 &= 1 - \gamma_| - x_2 \end{aligned}$$

In all similar argument forms the judgment of point probabilities reveals the "perceived correlations".

10 Second order probability distributions

The lower and upper bounds of the premise probabilities α_i span an n-dimensional hypercube with the volume $\prod_{i=1}^{n}(\alpha_i'' - \alpha_i')$. Coherence defines a subvolume within the hypercube. Each point in the subvolume corresponds to a vector of a coherent point probabilities. G-coherence requires the subvolume not to be empty. The ratio of the volume of the coherent volume and the total volume of the hypercube is a measure of the "degree of coherence" for a given pair of vectors of lower and upper probabilities.

If we treat the α_i as random variables, introduce rectangular density functions on the $[\alpha_i', \alpha_i'']$ intervals, $f(\alpha_i) = 1/(\alpha_i'' - \alpha_i')$, and if we assume that the α_i are stochastically independent, then volumes in the hypercube correspond to a probability measure. The volume of the coherent subspace measures the (second order) probability of being coherent.

It is however more general to replace the rectangular by more flexible distributions, to replace the intervals $[\alpha_i', \alpha_i'']$ by the full range of the unit interval $[0, 1]$, and to replace the independence assumption by an appropriate measure of probabilistic dependence [49]. The resulting structure is a *vine* structure [53]. It is characterized as follows:

1. The imprecise uncertainty of the n premises is modeled by a multivariate probability density on the simplex $[0, 1]^n$.

2. The (marginal) uncertainty of each premise is described by an appropriate probability density. For practical reasons we prefer to use beta distributions.

3. The pairwise (unconditional and conditional) stochastic dependencies are characterized by copulas. *Regular vines* allow a pairwise decomposition of the joint distributions.

4. Practical numerical analyses are performed by stochastic simulation.

The architecture corresponds to a *stochastic response model.* An individual represents his or her uncertainty by a distribution and when asked for a point probability judgment responds with a random number generated by the distribution. Inferences are performed by the propagation of second order probability distributions [49].

11 Discussion

All valid inference forms of propositional calculus which are conditional-free are p-valid. Of those containing conditionals a subset is p-nonvalid, most typically the PARADOXES OF THE MATERIAL IMPLICATION, STRENGTHENING THE ANTECEDENT (from $A \to C$ infer $(A \wedge B) \to C$), TRANSITIVITY (from from $A \to B$ and $B \to C$ infer $A \to C$), CONTRAPOSITION (from $A \to B$ infer $\neg B \to \neg A$), OR-TO-IF (from $A \vee B$ infer $\neg A \to B$). The p-nonvalid rules are just those which seem to be nonintuitive. TRANSITIVITY is an exception (it is probabilistically noninformative). For an extensive discussion based on the coherence approach see the recent paper on WEAK TRANSITIVITY by Gilio, Pfeifer, and Sanfilippo [40]. The conditional event interpretation is a filter that prevents nonintuitive rules to enter the inference system of probability logic. This is the reason why Adams used conditional probabilities for the probability of conditionals. Both, p-validity and conditional probabilities, go hand in hand.

Recent psychological studies [31, 86] used p-validity to evaluate human judgments as falling into "p-valid intervals". The intervals are claimed to be a new standard of rationality. These studies do not consider that Adams assigns interval probabilities with upper probabilities equal to 1 to the premises, that is, not point probabilities as in the judgments of the participants in the experiments. For inference rules like the MODUS PONENS or the MODUS TOLLENS, where all premises have degrees of essentialness equal to 1, Adams' uncertainty-sum criterion coincides with the lower probability of these rules when the conditionals are interpreted as material implication. If the inferences are not content-lean but involve every-day and causal knowledge the correlation between the events leads to conclusions with point probabilities. Moreover, psychologically it is even more plausible to represent the uncertain knowledge by continuous probability density functions. The strict classification as "coherent" and "non-coherent" dissolves and is replaced by distributions of degrees of coherence [49].

The coherence approach has a very elegant method to establish the bridge between classical logic and probability. In the coherence approach it is not necessary to start from a Boolean algebra. If the premises are *logically* dependent this is directly taken into account by removing impossible constituents, those that are forbidden by the logical dependence right at the beginning of any analysis [22].

For the treatment of probability preservation in generalized inference rules the reader is referred to [95], for degradation in probability logic to [94], and the for degradation in the context of exchangeable events to [93]. An inference form degrades if, as more and more premises are added to the premise set, the probability intervals of the conclusion get wider and wider. With the conditional event interpretation of conditionals the MODUS PONENS and the MODUS TOLLENS degrade so that even after a few steps the interval becomes probabilistically totally noninformative. The upper probabilities of the MODUS PONENS and the MODUS TOLLENS with material implication, $\gamma''_{\rightarrow} = \beta_{\rightarrow}$, depend on the probability of the conditionals only. Therefore for the material implication interpretation the upper bounds do not degrade as long as the probabilities of the conditionals remain constant.

Adams distinguished different kinds of probability preservation, among them certainty preservation [8]. "A is a *strict [certainty preserving] consequence* of S ... if and only if for all probability functions P ... if $P(B) = 1$ for all B in S, then $P(A) = 1$." [4, p. 274] McGee [60] observes that this criterion falls back to material implication: "The strictly valid inferences are not those described by Adams' theory, but those described by the orthodox theory, which treats the English conditional as the material conditional. This raises an ugly suspicion. The failures of the classical valid modes of inference appear only when we are reasoning from premises that are less than certain ... to a conclusion that is also less than certain." [60, p.189] This is a consequence of Adams' conception of conditional probability as defined by $P($ if A then $B)$ as $P(A \wedge B)/P(B)$ if $P(B) \neq 0$ and as 1 if $P(B) = 0$, i.e., he "...assigns the conditional the probability 1 when the conditional probability is undefined" [60, p. 190]. McGee proposed Popper functions, but zero probabilities are directly addressed in the coherence approach.

From a psychological perspective p-validity is attractive because it admits many rules that are endorsed by human reasoners and excludes logically valid rules that are not endorsed by human reasoners. People do not, for example, endorse the PARADOXES OF THE MATERIAL IMPLICATION [73]. This can, however, also be achieved with the criterion of *probabilistic noninformativeness* [70, 71, 73]. An inference form is noninformative if its premises do not constrain the probability of the conclusion. This is the case if the conclusion (represented by the vector of indicators) is linearly independent of the premises. In this case the probability of the conclusion is in the vacuous interval $[0, 1]$. P-nonvalid but informative inference forms, like DENYING THE ANTECEDENT or AFFIRMING THE CONSEQUENT, for example, should not be discredited as being "non-rational". They allow to constrain the probabilities of conclusions in the same way as the MODUS PONENS or the MODUS TOLLENS. In probabilistic inference coherence is the gold standard. In models of human reasoning p-validity is a fossil of classical logic.

References

[1] E. W. Adams. On rational betting systems (continuation). *Archiv für Mathematische Logik und Grundlagenforschung*, 6:112–128, 1961, 1964.

[2] E. W. Adams. On rational betting systems. *Archiv für Mathematische Logik und Grundlagenforschung*, 6:7–29, 1962.

[3] E. W. Adams. The logic of conditionals. *Inquiry: An Interdisciplinary Journal of Philosophy*, 8:166–197, 1965.

[4] E. W. Adams. Probability and the logic of conditionals. In J. Hintikka and P. Suppes, editors, *Aspects of Inductive Logic*, pages 265–316. North-Holland, Amsterdam, 1966.

[5] E. W. Adams. The logic of 'almost all'. *Journal of Philosophical Logic*, 3:3–17, 1974.

[6] E. W. Adams. *The Logic of Conditionals.* Reidel, Dordrecht, 1975.

[7] E. W. Adams. A note on comparing probabilistic and modal semantics of conditionals. *Theoria*, XLII(3):186–194, 1977.

[8] E. W. Adams. Four probability-preserving properties of inferences. *Journal of Philosophical Logic*, 25:1–24, 1996.

[9] E. W. Adams. *A Primer of Probability Logic.* CSLI Publications, Stanford, 1998.

[10] E. W. Adams and R. F. Fagot. A model of riskless choice. *Behavioral Choice*, 4:1–10, 1959.

[11] E. W. Adams and H. Levine. On the uncertainties transmitted from premises to conclusions in deductive inferences. *Synthese*, 30:429–460, 1975.

[12] T. Augustin, F.P. A. Colen, G. de Cooman, and M. C. M. Troffaes, editors. *Introduction to Imprecise Probabilities.* Wiley, Chichester, UK, 2014.

[13] D. Bamber. Entailment with near surety with scaled assertions of high probability. *Journal of Philosophical Logic*, 29:1–74, 2000.

[14] D. Bamber, I. R. Goodman, and H. T. Nguyen. High probability logic and inheritance. In J. W. Houpt and L. M. Blaha, editors, *Mathematical Models in Perception and Cognition. A Festschrift for James T. Tounsend*, volume I. Routletge, New York, 2016.

[15] S. Benferat, D. Dubois, and H. Prade. Nonmonotonic reasoning, conditional objects and possibility theory. *Artificial Intelligence*, 92:259–276, 1997.

[16] S. Benferat, D. Dubois, and H. Prade. Possibilistic and standard probabilistic semantics of conditional knowledge. In *AAAI-97 Proceedings*, pages 70–75. American Association for Artificial Intelligence, 1997.

[17] J. Bennett. *A Philosophical Guide to Conditionals.* Oxford University Press, Oxford, 2003.

[18] V. Biazzo and A. Gilio. A generalization of the Fundamental Theorem of de Finetti for imprecise conditional probability assessments. In *1st International Symposium on Imprecise Probabilities and Their Applications.* Electronic Version at http://decsai.ugr.es/ smc/isipta99/proc/009.html, Ghent, Belgium, 29 June–2 July 1999.

[19] V. Biazzo, A. Gilio, T. Lukasiewicz, and G. Sanfilippo. Probabilistic logic under coherence: complexity and algorithms. In *2nd International Symposium on Imprecise Probabilities and Their Applications.* Ithaca, New York, 2001.

[20] G. Boole. *An Investigation of the Laws of Thought.* Macmillan/Dover Publication, New York, 1854/1958.

[21] A. Capotorti, L. Galli, and B. Vantaggi. Locally strong coherence and inference with lower-upper probabililties. *Soft Computing*, 7:280–287, 2003.

[22] G. Coletti and R. Scozzafava. *Probabilistic Logic in a Coherent Setting.* Kluwer, Dordrecht, 2002.

[23] F. G. Cozman. Algorithms for conditioning on events of zero probability. In *Fifteenth International Florida Artificial Intelligence Society Conference (FLAIRS'02)*, pages 248–252. 2002.

[24] B. De Finetti. Foresight: Iits logical laws, its subjective sources (1937). In H. E. Jr. Kyburg and H. E. Smokler, editors, *Studies in Subjective Probability*, pages 93–158. Wiley, New York, 1964. orig. La prévsision: Ses lois logiques, ses sources subjectives. Annales de l'Institut Henri Poincaré, 7, 1-68.

[25] B. De Finetti. On the axiomatization of probability theory. In B. De Finetti, editor, *Probability, Induction and Statistics. The Art of Guessing*, pages 67–113. Wiley, London, 1972.

[26] B. De Finetti. *Theory of Probability. A critical Introductory Treatment.*, volume I. Wiley, London, 1974.

[27] L. E. Dubins. Finitely additive conditional probabilities, conglomerability and disintegrations. *The Annals of Probability*, 3:89–99, 1975.

[28] D. Dubois and H. Prade. Formal representation of uncertainty. In D. Bouyssou, D. Dubois, M. Pirlot, and H. Prade, editors, *Decision-Making Process.* ISTE/Wiley, London/Hoboken, N.J., 2009.

[29] J. St B. T. Evans, S. J. Handley, and D. E. Over. Conditionals and conditional probability. *Journal of Experimental Psychology: Learning, Memory, and Cognition*, 29:321–335, 2003.

[30] J. St B. T. Evans and D. E. Over. The probability of conditionals: The psychological evidence. *Mind and Language*, 18:340–358, 2003.

[31] J. St B. T. Evans, V. Thompson, and Over D. E. Uncertain deduction and conditional reasoning. *Frontiers in Psychology. Cognition*, 6:398, 2015.

[32] A. J. B. Fugard, N. Pfeifer, B. Mayerhofer, and G. D. Kleiter. How people interpret conditionals: Shifts toward the conditional event. *Journal of Experimental Psychology: Learning, Memory and Cognition*, 37:635–648, 2011.

[33] A. Gilio. Conditional events and subjective probability in management of uncertainty. In *Proceedings IPMU'92*, pages 165–168, Mallorca, 1992. Universitat de les Illes Balears.

[34] A. Gilio. Algorithms for precise and imprecise conditional probability assessments. In G. Coletti, D. Dubois, and R. Scozzafava, editors, *Mathematical Models for Handling Partial Knowledge in Artificial Intelligence*, pages 231–254. Planum Press, New York, 1995.

[35] A. Gilio. Probabilistic consistency of conditional probability bounds. In B. Bouchon Meunier, R. R. Yager, and L. A. Zadeh, editors, *Advances in Intelligent Computing (IPMU 94)*, volume 945 of *Lecture Notes in Computer Science*, pages 200–209. Springer, Berlin, 1995.

[36] A. Gilio. Probabilistic reasoning under coherence in system P. *Annals of Mathematics and Artificial Intelligence*, 34:5–34, 2002.

[37] A. Gilio. On Császár's condition in nonmonotonic reasoning. In *10th International Workshop on Non-Monotonic Reasoning. Special Session: Uncertainty Frameworks in Non-Monotonic Reasoning*, Whistler BC, Canada, June 6-8, 2004.

[38] A. Gilio and S. Ingrassia. Totally coherent set-values probability assessments. *Kybernetika*, 34:3–15, 1998.

[39] A. Gilio, D. E. Over, N. Pfeifer, and G. Sanfilippo. Centering and compound conditionals under coherence. In M. B. Ferraro, P. Giordani, B. Vantaggi, M. Gagolewski, M. A. Gil, P. Grzegorzewski, and O. Hryniewicz, editors, *Soft Methods for Data Science*, Advances in Intelligent Systems and Computing, pages 253–260. Springer, Dordrecht, 2017.

[40] A. Gilio, N. Pfeifer, and G. Sanfilippo. Transitivity in coherence-based probability logic. *Journal of Applied Logic*, 14:46–64, 2016.

[41] A. Gilio and G. Sanfilippo. Quasi conjunction and p-entailment in nonmonotonic reasoning. In C. Borgelt, G. González-Rodríguez, W. Trutschnig, M. A. Lubiano, M. A. Gil, P. Grzegorzewski, and O. Hryniewicz, editors, *Combining Soft Computing and Statistical Methods in Data Analysis*, volume 77 of *Advances in Intelligent and Soft Computing*, pages 321–328. Springer, Berlin, 2010.

[42] A. Gilio and G. Sanfilippo. Quasi conjunction and inclusion relation in probabilistic default reasoning. In W. Liu, editor, *Symbolic and Quantitative Approaches to Reasoning with Uncertainty, ESQUARU 2011*, volume 6717 of *Lecture Notes in Computer Science*, pages 497–508. Springer, Berlin, 2011.

[43] A. Gilio and G. Sanfilippo. Conjunction, disjunction and iterated conditioning of conditional events. In R. Kruse, M. R. Berthold, C. Moewes, M. Á. Gil, P. Grzegorzewski, and O. Hryniewicz, editors, *Synergies of Soft Computing and Statistics for Intelligent*

Data Analysis, volume 190 of *AISC*, pages 399–407. Springer, Berlin, Heidelberg, 2013. DOI: 10.1007/978-3-642-33042-1_43.

[44] I. R. Goodman, H. T. Nguyen, and E. A. Walker. *Conditional Inference and Logic for Intelligent Systems. A Theory of Measure-Free Conditioning.* North Holland, Amsterdam, 1991.

[45] Z. Gyenis, G. Hofer-Szabó, and M. Rédei. Conditioning using conditional expectations: The Borel-Kolmogorov paradox. *PhilSci Archive*, 2015.

[46] T. Hailperin. *Sentential Probability Logic.* Associated University Presses, Cranbury, NJ, beth edition, 1996.

[47] P. N. Johnson-Laird, S. S. Khemlani, and G. P. Goodwin. Logic, probability, and human reasoning. *Trends in Cognitive Sciences*, 19:201–214, 2015.

[48] G. D. Kleiter. *Bayes Statistik. Grundlagen und Anwendungen.* de Gruyter, Berlin, 1981.

[49] G. D. Kleiter. Modeling biased information seeking with second order probability distributions. *Kybernetika*, 51:469–485, 2015.

[50] G. D. Kleiter, A. J. B. Fugard, and N. Pfeifer. A process model of the understanding of uncertain conditionals. *Thinking and Reasoning*, in print, 2018.

[51] A. Kolmogoroff. *Grundbegriffe der Wahrscheinlichkeitsrechnung. Ergebnisse der Mathematik und ihrer Grenzgebiete.* Springer, Berlin, reprint 1973 edition, 1933.

[52] K. Kraus, D. Lehmann, and M. Magidor. Nonmonotonic reasoning, preferential models and cumulative logics. *Artificial Intelligence*, 44:167–207, 1990.

[53] D. Kurowicka and R. Joe. *Dependence Modeling: Vine Copula Handbook.* World Scientific, Singapure, 2011.

[54] F. Lad. *Operational Subjective Statistical Methods.* Wiley, New York, 1996.

[55] D. Lewis. *Counterfactuals.* Blackwell, Oxford, 1973.

[56] D. Lewis. Counterfactuals and comparative possibility. In W. L. Harper, R. Stalnaker, and G. Pearce, editors, *Ifs. Conditionals, Belief, Decision, Chance, and Time*, pages 57–85. Reidel, Dordrecht, NL, 1981.

[57] H. MacColl. *Symbolic Logic and its Applications.* Longmans, Green, and Co., New York, 1906. Reprint in "Forgotten Books", 2012.

[58] H. MacColl. 'If' and 'imply'. *Mind*, 17:151–152, 1908. Reprint in Rahman, S. and Redmond, J.: Hugh MacColl, An Overview of his Logical Work with Anthology. College Publications, London, 2007, p. 441-442.

[59] MATLAB. *version 7.10.0 (R2010a).* The MathWorks Inc., Natick, Massachusetts, 2012.

[60] V. McGee. Learning the impossible. In E. Eells and B. Skyrms, editors, *Probability and Conditionals*, pages 179–199. Cambridge University Press, Cambridge, UK, 1994.

[61] H. T. Nguyen and E. A. Walker. A history and introduction to the algebra of conditional events and probability logic. *IEEE Transactions on Systems, Man, and Cybernetics*, 24:1671–1675, 1994.

[62] M. Oaksford and N. Chater. *Bayesian Rationality. The Probabilistic Approach to Human Reasoning.* Oxford University Press, Oxford, 2007.

[63] M. Oaksford and N. Chater. Précis of Bayesian rationality: The probabilistic approach to human reasoning. *Behavioral and Brain Sciences*, 32:69–120, 2009.

[64] J. B. Paris and A. Vencovská. *Pure Inductive Logic.* Cambridge University Press, Cambridge, UK, 2015.

[65] J. Pearl. *Probabilistic Reasoning in Intelligent Systems: Networks of Plausible Inference.* Morgan Kaufmann Publishers, San Manteo, 1988.

[66] J. Pearl. From Adams' conditioinals to default expressions, causal conditionals, and counterfactuals. In E. Eells and B. Skyrms, editors, *Probability and Conditionals*, pages 47–74. Cambridge University Press, Cambridge, UK, 1994.

[67] N. Pfeifer. The new psychology of reasoning: A mental probability logical perspective. *Thinking and Reasoning*, 19:329–345, 2013.

[68] N. Pfeifer. Reasoning about uncertain conditionals. *Studia Logica*, 102:849–866, 2014.

[69] N. Pfeifer and G. D. Kleiter. Coherence and nonmonotonicity in human reasoning. *Synthese*, 146:93–109, 2005.

[70] N. Pfeifer and G. D. Kleiter. Inference in conditional probability logic. *Kybernetika*, 42:391–404, 2006.

[71] N. Pfeifer and G. D. Kleiter. Framing human inference by coherence based probability logic. *Journal of Applied Logic*, 7:206–217, 2009.

[72] N. Pfeifer and G. D. Kleiter. The conditional in mental probability logic. In M. Oaksford and N. Chater, editors, *Cognition and conditionals: Probability and logic in human thought*, pages 153–173. Oxford University Press, Oxford, 2010.

[73] N. Pfeifer and G. D. Kleiter. Uncertain deductive reasoning. In K. Manktelow, D. E. Over, and Elqayam S., editors, *The science of reasoning: A Festschrift for Jonathan St B.T. Evans*, pages 145–166. Psychology Press, Hove, UK, 2010.

[74] N. Pfeifer and G. D. Kleiter. Uncertainty in deductive reasoning. In K. Manktelow, D. Over, and S. Elqayam, editors, *The Science of Reason. A Festschrift for Jonathan St B. T. Evans*, pages 145–166. Psychology Press, Hove, UK, 2011.

[75] G. Politzer and J. Baratgin. Deductive schemas with uncertain premises using qualitative probability expressions. *Thinking and Reasoning*, 22:78–98, 2016.

[76] K. R. Popper. A set of independent axioms for probability. *Mind*, 47:275–277, 1938.

[77] K. R. Popper. *Logik der Forschung.* J. C. B. Mohr, Tübingen, 1976.

[78] E. Quaeghebeur. Desirability. In T. Augustin, F. P. A. Coolen, G. de Cooman, and M. C. M. Troffaes, editors, *Introduction to Imprecise Probabilities*, pages 1–27. Wiley, Chichester, UK, 2014.

[79] R Development Core Team. *R: A Language and Environment for Statistical Computing.* R Foundation for Statistical Computing, Vienna, Austria, 2009. ISBN 3-900051-07-0.

[80] M. J. Schervish, T. Seidenfeld, and J. B. Kadane. The extent of non-conglomerability of finitely additive probabilities. *Zeitschrift für Wahrscheinlichkeitstheorie und ver-*

wandte Gebiete, 66:205–226, 1984.

[81] G. Schurz. Probabilistic default logic based on irrelevance and relevance assumption. In Gabbay D., R. Kruse, A. Nonnegart, and H. J. Ohlbach, editors, *Qualitative and Quantitative Practical Reasoning*, number 1244 in Lecture Notes in Artificial Intelligence, pages 536–553. Springer, Berlin, 1997.

[82] G. Schurz. Probabilistic semantics for Delgrande's conditional logic and a counterexample to his default logic. *Artificial Intelligence*, 102:81–95, 1998.

[83] G. Schurz. Non-monotonic reasoning from an evolution-theoretic perspective: Ontic, logical and congnitive foundations. *Synthese*, 146:37–51, 2005.

[84] G. Schurz. *Wahrscheinlichkeit*. De Gruyter, Berlin, 2015.

[85] G. Schurz and P. D. Thorn. Reward versus risk in uncertain inference: Theorems and simulations. *The Review of Symbolic Logic*, 5:574–612, 2012.

[86] H. Singmann, K. C. Klauer, and D. E. Over. New normative standards of conditional reasoning and the dual-source model. *Frontiers in Psychology*, 5(PMC4029011):316, 2014.

[87] P. Suppes. Probabilistic inference and the concept of total evidence. In J. Hintikka and P. Suppes, editors, *Aspects of Inductive Logic*, pages 49–65. North-Holland, Amsterdam, 1966.

[88] P. Suppes. Some questions about Adams' conditionals. In E. Eells and B. Skyrms, editors, *Probability and Conditionals*, pages 5–11. Cambridge University Press, Cambridge, UK, 1994.

[89] C. G. Wagner. Modus tollens probabilized. *British Journal of Philosophy of Science*, 55:747–753, 2004.

[90] P. Walley. *Statistical Reasoning with Imprecise Probabilities.* Chapman and Hall, London, 1991.

[91] P. Walley. Measures of uncertainty in expert systems. *Artificial Intelligence*, 83:1–58, 1996.

[92] P. Walley, R. Pelessoni, and P. Vicig. Direct algorithms for checking coherence and making inferences from conditional probability assessments. Technical report, Quaderni del Dipartimento di Matematica Applicata alle Scienze Econoniche, Statistiche e Attuariali 'B. de Finetti' 6, 1999.

[93] C. Wallmann and G. D. Kleiter. Exchangeability in probability logic. In S. Greco, B. Bouchon-Meunier, G. Coletti, M. Fedrizzi, B. Matarazzo, and R. R. Yager, editors, *Proceedings of the 9th International Conference on Information Processing and Management of Uncertainty in Knowledge-Based Systems (IPMU)*, Communications in Computer and Information Science, Vol. 300, pages 157–167, Berlin, 2012. Springer.

[94] C. Wallmann and G. D. Kleiter. Degradation in probability logic: When more information leads to less precise conclusions. *Kybernetika*, 50:268–283, 2014.

[95] C. Wallmann and G.D. Kleiter. Probability propagation in generalized inference forms. *Studia Logica*, 102:913–929, 2014.

[96] K. Weichselberger. Axiomatic foundations of the theory of interval-probability. In

V. Mammitzsch and H. Schneeweiß, editors, *Symposia Gaussiana. Proceedings of the 2nd Gauss Symposium*, volume B, pages 47–64, Berlin, 1995. Walter de Gruyter.

[97] K. Weichselberger. *Elementare Grundbegriffe einer allgemeineren Wahrscheinlichkeitsrechnung I.* Physica-Verlag, Heidelberg, 2001.

[98] K. Weichselberger and T. Augustin. On the symbiosis of two concepts of conditional interval probability. In J.-M. Bernard, T. Seidenfeld, and M. Zaffalon, editors, *Proceedings of the Third International Symposium on Imprecise Probabilities and their Applications (ISIPTA'03)*, volume 18 of *Proceedings in Informatics*, pages 608–629. Carleton Scientific, Canada, 2003.

[99] K. Weichselberger and S. Pöhlmann. *A Methodology for Uncertainty in Knowledge-Based Systems.* Lecture Notes in Artificial Intelligence 419. Springer, Berlin, 1990.

[100] P. M. Williams. Notes on conditional previsions. *International Journal of Approximate Reasoning*, 44:366–383, 2007/1975. revised version of a 1975 Research Report, University of Sussex.

Received 5 July 2016

SUCCINCTNESS IN SUBSYSTEMS OF THE SPATIAL μ-CALCULUS

DAVID FERNÁNDEZ-DUQUE
Department of Mathematics, Ghent University
David.FernandezDuque@UGent.be

PETAR ILIEV
Institute de Recherche en Informatique de Toulouse, Toulouse University
petar.iliev@irit.fr

Abstract

In this paper we systematically explore questions of succinctness in modal logics employed in spatial reasoning. We show that the closure operator, despite being less expressive, is exponentially more succinct than the limit-point operator, and that the μ-calculus is exponentially more succinct than the equally-expressive tangled limit operator. These results hold for any class of spaces containing at least one crowded metric space or containing all spaces based on ordinals below ω^ω, with the usual limit operator. We also show that these results continue to hold even if we enrich the less succinct language with the universal modality.

1 Introduction

In spatial reasoning, as in any other application of logic, there are several criteria to take into account when choosing an appropriate formal system. A more expressive logic has greater potential applicability, but often at the cost of being less tractable. Similarly, a more *succinct* logic is preferable, for example, when storage capacity is limited: even when two formal languages $\mathcal{L}_1$ and $\mathcal{L}_2$ are equally expressive, it may be the case that certain properties are represented in $\mathcal{L}_1$ by much shorter expressions than in $\mathcal{L}_2$. As we will see, this is sometimes the case even when $\mathcal{L}_1$ is strictly *less*

This work was partially supported by ANR-11-LABX-0040-CIMI within the program ANR-11-IDEX-0002-02.

expressive than $\mathcal{L}_2$, and it may even be that $\mathcal{L}_1$ is exponentially more succinct than $\mathcal{L}_2$ and vice versa (for an intuitive explanation of this fact see [40] for example).

Qualitative spatial reasoning deals with regions in space and abstract relations between them, without requiring a precise description of them. It is useful in settings where data about such regions is incomplete or highly complex, yet precise numerical values of coordinates are not necessary: in such a context, qualitative descriptions may suffice and can be treated more efficiently from a computational perspective. One largely unexplored aspect of such efficiency lies in the succinctness of the formal languages employed. To this end, our goal is to study succinctness in the context of modal logics of space.

1.1 State-of-the-art in succinctness research

Succinctness is an important research topic that has been quite active for the last couple of decades. For example, it was shown by Grohe and Schweikardt [22] that the four-variable fragment of first-order logic is exponentially more succinct than the three-variable one on linear orders, while Eickmeyer et al. [10] offer a study of the succinctness of order-invariant sentences of first-order and monadic second-order logic on graphs of bounded tree-depth. Succinctness problems regarding temporal logics for formal verification of programs were studied, among others, by Wilke [41], Etessami et al. [11], and Adler and Immerman [2], while it was convincingly argued by Gogic et al. [18] that, as far as knowledge representations formalisms studied in the artificial intelligence are concerned, succinctness offers a more fine-grained comparison criterion than expressivity or computational complexity.

Intuitively, proving that one language $\mathcal{L}_1$ is more succinct than another language $\mathcal{L}_2$ ultimately boils down to proving a sufficiently big lower bound on the size of $\mathcal{L}_2$-formulas expressing some semantic property. For example, if we want to show that $\mathcal{L}_1$ is exponentially[1] more succinct than $\mathcal{L}_2$, we have to find an infinite sequence of semantic properties (i.e., classes of models) $\mathbf{P}_1, \mathbf{P}_2, \ldots$ definable in both $\mathcal{L}_1$ and $\mathcal{L}_2$, show that there are $\mathcal{L}_1$-formulas $\varphi_1, \varphi_2, \ldots$ defining $\mathbf{P}_1, \mathbf{P}_2, \ldots$ and prove that, for every n, every $\mathcal{L}_2$-formula ψ_n defining $\mathbf{P}_n$ has size exponential in the size of φ_n. Many such lower bound proofs, especially in the setting of temporal logics, rely on automata-theoretic arguments possibly combined with complexity-theoretic assumptions. In the present paper, we use formula-size games that were developed in the setting of Boolean function complexity by Razborov [35] and in the setting of first-order logic and some temporal logics by Adler and Immerman [2]. By now, the formula-size games have been adapted to a host of modal logics (see for example

[1] Analogous considerations apply in the case when we want to show that $\mathcal{L}_1$ is doubly exponentially, or non-elementarily, etc. more succinct than $\mathcal{L}_2$.

French et al. [16], Hella and Vilander [23], Figueira and Gorín [15], van der Hoek et al. [40]) and used to obtain lower bounds on modal formulas expressing properties of Kripke models. Our goal is to build upon these techniques in order to apply them to modal logics employed in spatial reasoning.

1.2 Spatial interpretations of modality

Modal logic is an extension of propositional logic with a 'modality' $\Diamond$ and its dual, $\Box$, so that if φ is any formula, $\Diamond\varphi$ and $\Box\varphi$ are formulas too. There are several interpretations for these modalities, but one of the first was studied by McKinsey and Tarski [30], who proposed a topological reading for them. These semantics have regained interest in the last decades due to their potential for spatial reasoning, especially when modal logic is augmented with a universal modality as studied by Shehtman [37], or fixpoint operators studied by Fernández-Duque [12] and Goldblatt and Hodkinson [21].

The intention is to interpret formulas of the modal language as subsets of a 'spatial' structure, such as $\mathbb{R}^n$. To do this, we use the *closure* and the *interior* of a set $A \subseteq \mathbb{R}^n$. The closure of A, denoted $c(A)$, is the set of points that have distance zero from A; its interior, denoted $i(A)$, is the set of points with positive distance from its complement. To define these, for $x, y \in \mathbb{R}^n$, let $\delta(x, y)$ denote the standard Euclidean distance between x and y. It is well-known that δ satisfies (i) $\delta(x, y) \geq 0$, (ii) $\delta(x, y) = 0$ iff $x = y$, (iii) $\delta(x, y) = \delta(y, x)$ and (iv) the triangle inequality, $\delta(x, z) \leq \delta(x, y) + \delta(y, z)$. More generally, a set X with a function $\delta: X \times X \to \mathbb{R}$ satisfying these four properties is a *metric space*. The Euclidean spaces $\mathbb{R}^n$ are metric spaces, but there are other important examples, such as the set of continuous functions on $[0, 1]$ (with a suitable metric).

Definition 1.1. *Given a metric space X and $A \subseteq X$, we say that a point x has* distance zero from A *if for every $\varepsilon > 0$, there is $y \in A$ so that $\delta(x, y) < \varepsilon$. If x does not have distance zero from A, we say it has* positive distance *from A. Then, $c(A)$ is the set of points with zero distance from A and $i(A)$ is the set of points with positive distance from its complement.*

Note that if we denote the complement of A by $\overline{A}$, then $i(A) = \overline{c(\overline{A})}$. The basic properties of c are well-known, and we mention them without proof.

Proposition 1.2. *If X is a metric space and c is the closure operator on X, then, given sets $A, B \subseteq X$,*

(i) $c(\varnothing) = \varnothing$,

(ii) $A \subseteq c(A)$,

(iii) $c(A) = c(c(A))$ *and*

(iv) $c(A \cup B) = c(A) \cup c(B)$.

These four properties are known as the *Kuratowski axioms* [28]. We will say that any non-empty set X equipped with a function $c: 2^X \to 2^X$ satisfying the Kuratowski axioms is a *closure space,* and that c is a *closure operator.* Closure spaces are simply topological spaces in disguise, but presenting them in this fashion will have many advantages for us. To be precise, if (X, c) is a closure space and $A = c(A)$, we say that X is *closed,* and its complement is *open;* the family of open sets then gives a topology in the usual way.[2]

From a computational perspective, it can be more convenient to work with closure spaces than with metric spaces, as finite, non-trivial closure spaces can be defined in a straightforward way, and thus spatial relations can be represented using finite structures. To be precise, let W be a non-empty set and $R \subseteq W \times W$ a binary relation; the structure (W, R) is a *frame.* Then, if R is a preorder (i.e., a transitive, reflexive relation), the operator $R^{-1}[\cdot]$ defined by $R^{-1}[A] = \{w \in W : \exists v \in A\ (w\ R\ v)\}$ is a closure operator.

A good deal of the geometric properties of regions in a metric space X are reflected in the behavior of its closure operator; however, some information is inevitably lost. It has been observed that more information about the structure of X is captured if we instead consider its *limit,* or *set-derivative,* operator. For $A \subseteq X$, define $d(A)$ to be the set of points such that, for every $\varepsilon > 0$, there is $y \in A$ *different* from x such that $\delta(x, y) < \varepsilon$. The limit operator was first considered in the modal logic literature by McKinsey and Tarski [30] and has since been extensively studied (see e.g. [4, 21, 29]).

Note that it is no longer the case that $A \subseteq d(A)$: for example, if $A = \{x\}$ consists of a single point, then $d(A) = \varnothing$. Nevertheless, d still satisfies the following properties.

Proposition 1.3. *Let X be a metric space, and let $d: 2^X \to 2^X$ be its limit operator. Then, for any $A, B \subseteq X$,*

(i) $d(\varnothing) = \varnothing$,

(ii) $d(d(A)) \subseteq d(A)$,[3] *and*

[2] We will not define topological spaces in this text, and instead refer the reader to a text such as [32].

[3] If, instead, we let X be an arbitrary topological space, then only the weaker condition $d(d(A)) \subseteq d(A) \cup A$ holds in general.

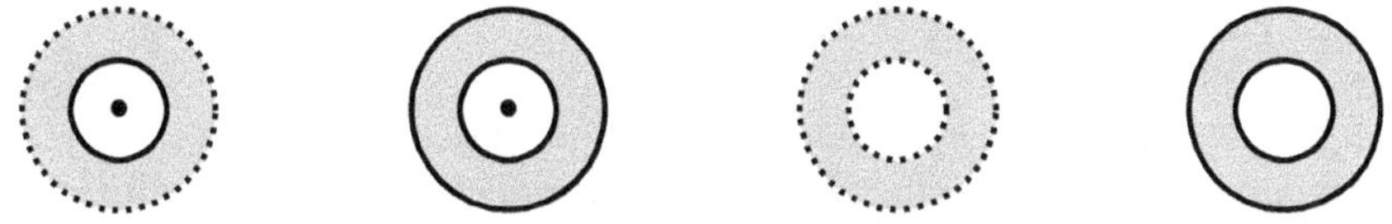

Figure 1: A region in $\mathbb{R}^2$ and its closure, interior and limit sets. The point in the middle is the only isolated point of A.

(iii) $d(A \cup B) = d(A) \cup d(B)$.

In order to treat closure operators and limit operators uniformly, we will define a *convergence space* to be a pair (X, d), where $d: 2^X \to 2^X$ satisfies these three properties (see Definition 2.2). In any convergence space, we can then define $c(A) = A \cup d(A)$, but in general, it is not possible to define d in terms of c using Boolean operations. In particular, the *isolated points* of A can be defined as the elements of $A \setminus d(A)$, but they cannot be defined in terms of c (see Figure 1). As before, a convenient source of convergence spaces is provided by finite frames: if (W, R) is such that R is transitive (but not necessarily reflexive), then $(W, R^{-1}[\cdot])$ is a convergence space.

Logics involving the d operator are more expressive than those with c alone, for example being able to distinguish the real line from higher-dimensional space and Euclidean spaces from arbitrary metric spaces, as shown by Bezhanishvili et al. [4]. Nevertheless, as we will see, the modal language with the c-interpretation is exponentially more succinct than modal logic with the d-interpretation.

Layout of the article

In Section 2, we review the syntax and semantics of the spatial μ-calculus, extended in Section 3 to include the closure and tangled limit operators. Section 4 then reviews some model-theoretic operations that preserve the truth of modal formulas, and Section 5 discusses the classes of spaces that we will be interested in.

Section 6 presents the game-theoretic techniques that we will use to establish our core results in Section 7, where it is shown that the closure operator is more succinct than the limit-point operator. This is extended to include the tangled limit operator and the universal modality in Section 8, leading to our main results, namely Theorem 8.6 and Theorem 8.7. Finally, Section 9 provides some concluding remarks and open problems.

2 The spatial μ-calculus

In this section we present the modal μ-calculus and formalize its semantics over *convergence spaces*, a general class of models that allow us to study all the semantic structures we are interested in under a unified framework. Let us begin by defining the basic formal language we will work with.

2.1 Syntax

We will consider logics over variants of the language $\mathcal{L}^{\mu}_{\Diamond\forall}$ given by the following grammar (in Backus-Naur form). Fix a set P of *propositional variables* (also called *atoms*), and define:

$$\varphi, \psi := \top \mid \bot \mid p \mid \overline{p} \mid \varphi \vee \psi \mid \varphi \wedge \psi \mid \Diamond\varphi \mid \Box\varphi \mid \forall\varphi \mid \exists\varphi \mid \mu p.\varphi \mid \nu p.\varphi$$

Here, $p \in P$, $\overline{p}$ denotes the negation of p, and $\overline{p}$ may not occur in $\mu p.\varphi$ or $\nu p.\varphi$. For the game-theoretic techniques we will use, it is convenient to allow negations only at the atomic level, and thus we include all duals as primitives, but not negation or implication; however, we may use the latter as shorthands, defined via De Morgan's laws. As usual, formulas of the forms $p, \overline{p}$ are *literals*. It will also be crucial for our purposes to measure the size of formulas: the *size* of a formula φ is denoted $|\varphi|$ and is defined as the number of nodes in its syntax tree.

Definition 2.1. *We define a function* $|\cdot| : \mathcal{L}^{\mu}_{\Diamond\forall} \to \mathbb{N}$ *recursively by*

- $|p| = |\overline{p}| = 1$
- $|\varphi \wedge \psi| = |\varphi \vee \psi| = |\varphi| + |\psi| + 1$
- $|\Diamond\varphi| = |\Box\varphi| = |\forall\varphi| = |\exists\varphi| = |\mu p.\varphi| = |\nu p.\varphi| = |\varphi| + 1.$

Sublanguages of $\mathcal{L}^{\mu}_{\Diamond\forall}$ are denoted by omitting some of the operators, with the convention that whenever an operator is included, so is its dual (for example, $\mathcal{L}_{\Diamond}$ includes the modalities $\Diamond, \Box$ but does not allow $\forall, \exists, \mu$ or ν).

2.2 Convergence spaces

Spatial interpretations of modal logics are usually presented in terms of topological spaces. Here we will follow an unorthodox route, instead introducing *modal spaces,* and as a special case *convergence spaces;* both the closure and limit point operators give rise to convergence spaces, while Kripke frames can be seen as modal spaces. As a general convention, structures (e.g. frames or models) will be denoted $\mathcal{A}, \mathcal{B}, \ldots,$

while classes of structures will be denoted **A**, **B**, ... The domain of a structure $\mathcal{A}$ will be denoted by $|\mathcal{A}|$.

Definition 2.2. *A* normal operator *on a set A is any function $\rho: 2^A \to 2^A$ satisfying*

(1) $\rho(\varnothing) = \varnothing$, and

(2) $\rho(X \cup Y) = \rho(X) \cup \rho(Y)$.

A modal space *is a pair $\mathcal{A} = (|\mathcal{A}|, \rho_{\mathcal{A}})$, where $|\mathcal{A}|$ is any non-empty set and $\rho_{\mathcal{A}}: 2^{|\mathcal{A}|} \to 2^{|\mathcal{A}|}$ is a normal operator. If $X \subseteq |\mathcal{A}|$, define $\hat{\rho}_{\mathcal{A}}(X) = \overline{\rho_{\mathcal{A}}(\overline{X})}$. We say that $\rho_{\mathcal{A}}$ is a* convergence operator *if it also satisfies*

(3) $\rho_{\mathcal{A}}(\rho_{\mathcal{A}}(X)) \subseteq \rho_{\mathcal{A}}(X)$.

If $\rho_{\mathcal{A}}$ is a convergence operator, we will say that $\mathcal{A}$ is a convergence space.

If $X \subseteq \rho_{\mathcal{A}}(X)$ for all $X \subseteq |\mathcal{A}|$, we say that $\rho_{\mathcal{A}}$ is inflationary. *An inflationary convergence operator is a* closure operator, *and if $\rho_{\mathcal{A}}$ is a closure operator then $\mathcal{A}$ is a* closure space.

This general presentation will allow us to unify semantics over metric spaces with those over arbitrary relational structures.

Remark 2.3. *Modal spaces are just* neighborhood structures *[33] in disguise; indeed, if (A, N) is a neighborhood structure (i.e., $N \subseteq A \times 2^A$), we may set $\rho_N(X)$ to be the set of $a \in A$ such that every neighborhood of a intersects X. Conversely, if $X \subseteq |\mathcal{A}|$ and $a \in |\mathcal{A}|$, we let X be a neighborhood of a if $a \in \hat{\rho}_{\mathcal{A}}(X)$.*

As mentioned before, a Kripke frame is simply a structure $\mathcal{A} = (|\mathcal{A}|, R_{\mathcal{A}})$, where $|\mathcal{A}|$ is a non-empty set and $R_{\mathcal{A}} \subseteq |\mathcal{A}| \times |\mathcal{A}|$ is a binary relation. We will implicitly identify $\mathcal{A}$ with the modal space $(|\mathcal{A}|, \rho_{\mathcal{A}})$, where $\rho_{\mathcal{A}}(X) = R_{\mathcal{A}}^{-1}[X]$ for any $X \subseteq |\mathcal{A}|$. It is readily verified that the operator $\rho_{\mathcal{A}}$ thus defined is normal. In this sense, modal spaces generalize Kripke frames, but the structure $(|\mathcal{A}|, \rho_{\mathcal{A}})$ is not always a convergence space. Nevertheless, convergence spaces may be obtained by restricting our attention to frames where $R_{\mathcal{A}}$ is transitive.

Definition 2.4. *Define* **K** *to be the class of all Kripke frames,* **K4** *to be the class of all Kripke frames with a transitive relation,* **KD4** *to be the class of all Kripke frames with a transitive, serial*[4] *relation, and* **S4** *to be the class of all Kripke frames with a transitive, reflexive relation.*

[4] Recall that a relation $R \subseteq W \times W$ is *serial* if $R(w) \neq \varnothing$ for all $w \in W$.

The names for these classes are derived from their corresponding modal logics (see e.g. [8]). The following is then readily verified, and we mention it without proof:

Proposition 2.5.

1. *If $\mathcal{A} \in \mathbf{K}$, then $(|\mathcal{A}|, R_{\mathcal{A}}^{-1}[\cdot])$ is a modal space.*

2. *If $\mathcal{A} \in \mathbf{K4}$, then $(|\mathcal{A}|, R_{\mathcal{A}}^{-1}[\cdot])$ is a convergence space.*

3. *If $\mathcal{A} \in \mathbf{S4}$, then $(|\mathcal{A}|, R_{\mathcal{A}}^{-1}[\cdot])$ is a closure space.*

Before defining our semantics, we need to briefly discuss least and greatest fixed points. Let X be a set and $f: 2^X \to 2^X$ be monotone; that is, if $A \subseteq B \subseteq X$, then $f(A) \subseteq f(B)$. Say that A_* is a *fixed point* of f if $f(A_*) = A_*$. Then, Knaster and Tarski [39] showed the following:

Theorem 2.6. *If X is any set and $f: 2^X \to 2^X$ is monotone, then*

1. *f has a $\subseteq$-least fixed point, which we denote* $\mathrm{LFP}(f)$, *and*

2. *f has a $\subseteq$-greatest fixed point, which we denote* $\mathrm{GFP}(f)$.

This result is discussed in some detail in the context of the spatial μ-calculus in [21]. With this, we turn our attention from frames to *models*.

2.3 Models and truth definitions

Formulas of $\mathcal{L}^{\mu}_{\Diamond\forall}$ are interpreted as subsets of a convergence space, but first we need to determine the propositional variables that are true at each point.

Definition 2.7. *If $\mathcal{A}$ is a modal space, a* valuation *on $\mathcal{A}$ is a function $V: |\mathcal{A}| \to 2^P$ (recall that P is the set of atoms). A modal space $\mathcal{A}$ equipped with a valuation V is a* model. *If $\mathcal{A}$ is a convergence space, then $(\mathcal{A}, V)$ is a* convergence model.

If $X \subseteq |\mathcal{A}|$ and $p \in P$, we define a new valuation $V_{[X/p]}$ by setting

- for $q \neq p$, $q \in V_{[X/p]}(w)$ if and only if $q \in V(w)$, and
- $p \in V_{[X/p]}(w)$ if and only if $w \in X$.

Now we are ready to define the semantics for $\mathcal{L}^{\mu}_{\Diamond\forall}$.

Definition 2.8. *Let $(\mathcal{A}, V)$ be a model. We define the truth set*

$$[\![\varphi]\!]_V = \{w \in |\mathcal{A}| : (\mathcal{A}, w) \models \varphi\}$$

by structural induction on φ. We first need an auxiliary definition for the cases $\mu p.\varphi$ and $\nu p.\varphi$. Suppose inductively that $[\![\varphi]\!]_{V'}$ has been defined for any valuation V', and define a function $\lambda p.\varphi[V]: 2^{|\mathcal{A}|} \to 2^{|\mathcal{A}|}$ given by $\lambda p.\varphi[V](X) = [\![\varphi]\!]_{V_{[X/p]}}$. With this, we define:

$$\begin{array}{lcllcl}
[\![p]\!]_V & = & \{w \in |\mathcal{A}| : p \in V(w)\} & [\![\overline{p}]\!]_V & = & \{w \in |\mathcal{A}| : p \notin V(w)\} \\
[\![\varphi \wedge \psi]\!]_V & = & [\![\varphi]\!]_V \cap [\![\psi]\!]_V & [\![\varphi \vee \psi]\!]_V & = & [\![\varphi]\!]_V \cup [\![\psi]\!]_V \\
[\![\Diamond\varphi]\!]_V & = & \rho_{\mathcal{A}}([\![\varphi]\!]_V) & [\![\Box\varphi]\!]_V & = & \hat{\rho}_{\mathcal{A}}([\![\varphi]\!]_V) \\
[\![\exists\varphi]\!]_V & = & X \textit{ if } [\![\varphi]\!]_V \neq \varnothing \textit{ else } \varnothing & [\![\forall\varphi]\!]_V & = & X \textit{ if } [\![\varphi]\!]_V = X \textit{ else } \varnothing \\
[\![\mu p.\varphi]\!]_V & = & \mathrm{LFP}(\lambda p.\varphi[V]) & [\![\nu p.\varphi]\!]_V & = & \mathrm{GFP}(\lambda p.\varphi[V]).
\end{array}$$

Given a model $(\mathcal{A}, V)$ and formulas $\varphi, \psi \in \mathcal{L}^{\mu}_{\Diamond\forall}$, we say that φ is equivalent to ψ *on $\mathcal{A}$ if $[\![\varphi]\!]_V = [\![\psi]\!]_V$. If $\mathcal{A}$ is a modal space, φ, ψ are equivalent on $\mathcal{A}$ if they are equivalent on any model of the form $(\mathcal{A}, V)$, and if $\mathbf{A}$ is a class of structures, we say that φ, ψ are equivalent over $\mathbf{A}$ if they are equivalent on any element of $\mathbf{A}$. When $\mathcal{A}$ or $\mathbf{A}$ is clear from context, we may write $\varphi \equiv \psi$, and if $\varphi \equiv \top$ we say φ is* valid, *in which case we write $\mathcal{A} \models \varphi$ or $\mathbf{A} \models \varphi$, respectively.*

It is readily verified that if $\overline{p}$ does not appear in φ, it follows that $\lambda p.\varphi[V]$ is monotone, and hence the above definition is sound. On occasion, if $\mathcal{M}$ is a model with valuation V, we may write $[\![\cdot]\!]_{\mathcal{M}}$ or even $[\![\cdot]\!]$ instead of $[\![\cdot]\!]_V$. In the case that $(\mathcal{A}, V)$ is a Kripke model, the semantics we have just defined coincide with the standard relational semantics [8]. To see this, note that for any formula φ and any $w \in |\mathcal{A}|$, $w \in [\![\Diamond\varphi]\!]$ if and only if $w \in \rho_{\mathcal{A}}([\![\varphi]\!])$, which means that $w \in R^{-1}_{\mathcal{A}}[[\![\varphi]\!]]$; i.e., there is $v \in [\![\varphi]\!]$ such that $w \; R_{\mathcal{A}} \; v$. Thus, the interpretation of $\Diamond$ coincides with the standard relational interpretation in modal logic.

3 The extended spatial language

The spatial μ-calculus, as we have presented it, may be naturally extended to include other definable operations. Of course such extensions will not add any expressive power to our language, but as we will see later in the text, they can yield considerable gains in terms of succinctness. We begin by discussing the closure operator.

3.1 The closure operator

As we have mentioned, the closure operator is definable in terms of the limit point operator on metric spaces. Let us make this precise. We will denote the closure operator by $\oplus\varphi$, defined as a shorthand for $\varphi \vee \Diamond\varphi$. Dually, the interior operator $\boxplus\varphi$ will be defined as $\varphi \wedge \Box\varphi$. To do this, let $\mathcal{L}^{\mu}_{\oplus\Diamond\forall}$ be the extension of $\mathcal{L}^{\mu}_{\Diamond\forall}$ which includes $\oplus, \boxplus$ as primitives. Then, for $\varphi \in \mathcal{L}^{\mu}_{\oplus\Diamond\forall}$, define a formula[5] $t^{\Diamond}_{\oplus}(\varphi) \in \mathcal{L}^{\mu}_{\Diamond\forall}$ by letting $t^{\Diamond}_{\oplus}(\cdot)$ commute with Booleans and all modalities except for $\oplus, \boxplus$, in which case $t^{\Diamond}_{\oplus}(\oplus\varphi) = t^{\Diamond}_{\oplus}(\varphi) \vee \Diamond t^{\Diamond}_{\oplus}(\varphi)$ and $t^{\Diamond}_{\oplus}(\boxplus\varphi) = t^{\Diamond}_{\oplus}(\varphi) \wedge \Box t^{\Diamond}_{\oplus}(\varphi)$.

Semantics for $\mathcal{L}^{\mu}_{\oplus\Diamond\forall}$ are defined by setting $[\![\varphi]\!]_V = [\![t^{\Diamond}_{\oplus}(\varphi)]\!]_V$, and we extend Definition 2.1 to $\mathcal{L}^{\mu}_{\oplus\Diamond\forall}$ in the obvious way, by

$$|\oplus\varphi| = |\boxplus\varphi| = |\varphi| + 1.$$

With this, we can give an easy upper bound on the translation $t^{\Diamond}_{\oplus}$.

Lemma 3.1. *If $\varphi \in \mathcal{L}^{\mu}_{\oplus\Diamond\forall}$, then $\left|t^{\Diamond}_{\oplus}(\varphi)\right| \leq 2^{|\varphi|}$.*

However, this bound is not optimal; it can be improved if we instead define $\Diamond$ in terms of μ. Define $t^{\mu}_{\oplus}\colon \mathcal{L}^{\mu}_{\oplus\Diamond\forall} \to \mathcal{L}^{\mu}_{\Diamond\forall}$ by replacing every occurrence of $\oplus\varphi$ recursively by $\mu p.(t^{\mu}_{\oplus}(\varphi) \vee \Diamond p)$ and every occurrence of $\boxplus\varphi$ recursively by $\nu p.(t^{\mu}_{\oplus}(\varphi) \wedge \Box p)$ (where p is always a fresh variable), and commuting with Booleans and other operators. Then, we obtain the following:

Lemma 3.2. *For all $\varphi \in \mathcal{L}^{\mu}_{\oplus\Diamond\forall}$, we have that $\varphi \equiv t^{\mu}_{\oplus}(\varphi)$ over the class of convergence spaces, and $\left|t^{\mu}_{\oplus}(\varphi)\right| \leq 4|\varphi|$.*

We omit the proof, which is straightforward. Whenever $\Diamond$ is interpreted as a convergence operator, $\oplus$ is then interpreted as a closure operator. To be precise, given a modal space $\mathcal{A}$, define a new operator $\rho^{+}_{\mathcal{A}}$ on $|\mathcal{A}|$ by $\rho^{+}_{\mathcal{A}}(X) = X \cup \rho_{\mathcal{A}}(X)$. Then, $\mathcal{A}^{+} = (|\mathcal{A}|, \rho^{+}_{\mathcal{A}})$ is an inflationary modal space, and if $\mathcal{A}$ is a convergence space, it follows that $\mathcal{A}^{+}$ is a closure space. If $(\mathcal{A}, V)$ is any model and φ is any formula, it is straightforward to check that $[\![\oplus\varphi]\!]_V = \rho^{+}_{\mathcal{A}}([\![\varphi]\!]_V)$. In the setting of Kripke models, we see that $w \in [\![\oplus\varphi]\!]_V$ if and only if $w \in [\![\varphi]\!]_V$, or there is $v \in W$ such that $wR_{\mathcal{A}}v$ and $v \in [\![\varphi]\!]_V$; as was the case for $\Diamond\varphi$, this coincides with the standard relational semantics, but with respect to the reflexive closure of $R_{\mathcal{A}}$.

3.2 Tangled limits

There is one final extension to our language of interest to us; namely, the tangled limit operator, known to be expressively equivalent to the μ-calculus over the class of convergence spaces, but with arguably simpler syntax and semantics.

[5]We use the general convention that the symbol being replaced by a translation is placed as a subindex, and the symbol used to replace it is used as a superindex. However, this convention is only orientative.

Definition 3.3. *Let $\mathcal{L}^{\mu\Diamond^*}_{\oplus\Diamond\forall}$ be the extension of $\mathcal{L}^{\mu}_{\oplus\Diamond\forall}$ such that, if $\varphi_1,\ldots,\varphi_n$ are formulas, then so are $\Diamond^*\{\varphi_1,\ldots,\varphi_n\}$ and $\Box^*\{\varphi_1,\ldots,\varphi_n\}$. We define $t^{\mu}_{\Diamond^*}(\varphi)$ to commute with all operators except $\Diamond^*,\Box^*$, in which case*

$$t^{\mu}_{\Diamond^*}(\Diamond^*\{\varphi_1,\ldots,\varphi_n\}) = \mu p. \bigwedge_{i\leq n} \Diamond(p \wedge t^{\mu}_{\Diamond^*}(\varphi_i))$$
$$t^{\mu}_{\Diamond^*}(\Box^*\{\varphi_1,\ldots,\varphi_n\}) = \nu p. \bigvee_{i\leq n} \Box(p \vee t^{\mu}_{\Diamond^*}(\varphi_i)).$$

We extend $|\cdot|$ and $[\![\cdot]\!]$ to $\mathcal{L}^{\mu\Diamond^*}_{\oplus\Diamond\forall}$ by defining

$$|\Diamond^*\{\varphi_1,\ldots,\varphi_n\}| = |\Box^*\{\varphi_1,\ldots,\varphi_n\}| = |\varphi_1| + \ldots + |\varphi_n| + 1$$

and $[\![\varphi]\!]_V = \left[\!\left[t^{\mu}_{\Diamond^*}(\varphi)\right]\!\right]_V$.

We call $\Diamond^*$ the *tangled limit operator;* this was introduced by Dawar and Otto [9] in the context of **K4** frames, then extended by Fernández-Duque [12] to closure spaces and by Goldblatt and Hodkinson [21] to other convergence spaces. For clarity, let us give a direct definition of $\Diamond^*$ without translating into the μ-calculus.

Lemma 3.4. *If $(\mathcal{A},V)$ is a convergence model, $\varphi_1,\ldots,\varphi_n$ any sequence of formulas, and $x \in |\mathcal{A}|$, then $x \in [\![\Diamond^*\{\varphi_1,\ldots,\varphi_n\}]\!]_V$ if and only if there is $S \subseteq |\mathcal{A}|$ such that $x \in S$ and, for all $i \leq n$, $S \subseteq \rho_{\mathcal{A}}(S \cap [\![\varphi_i]\!]_V)$.*

Although Dawar and Otto proved in [9] that the tangled limit operator is equally expressive as the μ-calculus, they use model-theoretic techniques that do not provide an explicit translation. As such, we do not provide an upper bound in the following result.

Theorem 3.5. *There exists a function $t^{\Diamond^*}_{\mu}:\mathcal{L}^{\mu}_{\Diamond} \to \mathcal{L}^{\Diamond^*}_{\Diamond}$ such that, for all $\varphi \in \mathcal{L}^{\mu}_{\Diamond}$, $\varphi \equiv t^{\Diamond^*}_{\mu}(\varphi)$ on the class of* **K4** *frames.*

Spatial interpretations of the tangled closure and limit operators have gathered attention in recent years (see e.g. [13, 14, 19, 20]). Later we will show that the tangled limit operator, despite being equally expressive, is exponentially less succinct than the μ-calculus.

4 Truth-preserving transformations

Let us review some notions from the model theory of modal logics, and lift them to the setting of convergence spaces. We begin by discussing bisimulations, the standard notion of equivalence between Kripke models; or, more precisely between *pointed models,* which are pairs $(\mathcal{A},a)$ such that $\mathcal{A}$ is a model and $a \in |\mathcal{A}|$.

4.1 Bisimulations

The well-known notion of *bisimulation* between Kripke models readily generalizes to the setting of convergence spaces, using what we call *confluent* relations. Below, we say that two pointed models $(\mathcal{A}, a)$ and $(\mathcal{B}, b)$ *differ* on the truth of a propositional variable p when we have $(\mathcal{A}, a) \models p$ whereas $(\mathcal{B}, b) \models \overline{p}$, or vice-versa. If $(\mathcal{A}, a)$ and $(\mathcal{B}, b)$ do not differ on p, then they *agree* on p.

Definition 4.1. *Let $\mathcal{A} = (|\mathcal{A}|, \rho_{\mathcal{A}})$ and $\mathcal{B} = (|\mathcal{B}|, \rho_{\mathcal{B}})$ be modal spaces and $\chi \subseteq |\mathcal{A}| \times |\mathcal{B}|$. We say that χ is* forward confluent *if, for all $X \subseteq |\mathcal{A}|$,*

$$\chi[\rho_{\mathcal{A}}(X)] \subseteq \rho_{\mathcal{B}}(\chi[X]).$$

Say that χ is backward confluent *if χ^{-1} is forward confluent, and* confluent *if it is forward and backward confluent.*

Let $Q \subseteq P$ be a set of atoms. If $(\mathcal{A}, V_{\mathcal{A}})$ and $(\mathcal{B}, V_{\mathcal{B}})$ are models, a bisimulation relative to Q *is a confluent relation $\chi \subseteq |\mathcal{A}| \times |\mathcal{B}|$ such that if $a \ \chi \ b$, then $(\mathcal{A}, a)$ agrees with $(\mathcal{B}, b)$ on all atoms of Q. If Q is not specified, we assume that $Q = P$.*

A bisimulation between pointed models $(\mathcal{A}, a)$ and $(\mathcal{B}, b)$ is a bisimulation $\chi \subseteq |\mathcal{A}| \times |\mathcal{B}|$ such that $a \ \chi \ b$. We say that $(\mathcal{A}, a)$ and $(\mathcal{B}, b)$ are locally bisimilar *if there exists a bisimulation between them, in which case we write $(\mathcal{A}, a) \leftrightarroweq (\mathcal{B}, b)$. They are* globally bisimilar *if there exists a total, surjective bisimulation between them.*

The following is readily verified by a structural induction on formulas (see, for example, [21]). Recall that a variable p is *free* if it appears outside of the scope of μp or νp.

Lemma 4.2. *Let $(\mathcal{A}, a)$ and $(\mathcal{B}, b)$ be pointed models, $Q \subseteq P$, and $\varphi \in \mathcal{L}^{\mu\Diamond^*}_{\oplus\Diamond\forall}$, all of whose free atoms appear in Q. Then, if either*

1. *$(\mathcal{A}, a)$ and $(\mathcal{B}, b)$ are locally bisimilar relative to Q and $\varphi \in \mathcal{L}^{\mu\Diamond^*}_{\oplus\Diamond}$ (i.e., $\forall, \exists$ do not appear in φ), or*

2. *$(\mathcal{A}, a)$ and $(\mathcal{B}, b)$ are globally bisimilar relative to Q,*

it follows that $(\mathcal{A}, a) \models \varphi$ if and only if $(\mathcal{B}, b) \models \varphi$.

As a special case, we can view an *isomorphism* as a bisimulation that is also a bijection. Isomorphism between structures will be denoted by $\cong$. Similarly, if $|\mathcal{A}| = |\mathcal{B}|$ and Q is a set of atoms, we say that $\mathcal{A}$ and $\mathcal{B}$ *agree on all atoms in Q* if for every $w \in |\mathcal{A}|$, $(\mathcal{A}, w)$ agrees with $(\mathcal{B}, w)$ on all atoms in Q. It is easy to see that if $\mathcal{A}$ and $\mathcal{B}$ agree on all atoms in Q, then $\mathcal{A}$ and $\mathcal{B}$ are globally bisimilar relative to Q.

It is instructive to compare our notion of confluence to more familiar notions in the literature. We begin with the familiar notion of bisimulations between relational models:

Lemma 4.3. *If $\mathcal{A} = (|\mathcal{A}|, R_{\mathcal{A}})$ and $\mathcal{B} = (|\mathcal{B}|, R_{\mathcal{B}})$ are Kripke frames, then $\chi \subseteq |\mathcal{A}| \times |\mathcal{B}|$ is forward-confluent if and only if, whenever $a \, R_{\mathcal{A}} \, a'$ and $a \, \chi \, b$, there is $b' \in |\mathcal{B}|$ such that $a' \, \chi \, b'$ and $b \, R_{\mathcal{B}} \, b'$.*

On metric spaces, confluent functions are related to continuous and open maps.

Lemma 4.4. *Let $\mathcal{A} = (|\mathcal{A}|, \delta_{\mathcal{A}})$ and $\mathcal{B} = (|\mathcal{B}|, \delta_{\mathcal{B}})$ be metric spaces with respective closure operators $c_{\mathcal{A}}$, $c_{\mathcal{B}}$ and limit operators $d_{\mathcal{A}}, d_{\mathcal{B}}$, and let $f: |\mathcal{A}| \to |\mathcal{B}|$. Then,*

1. *f is forward-confluent with respect to $c_{\mathcal{A}}$, $c_{\mathcal{B}}$ if and only if f is* continuous; *that is, for every $a \in |\mathcal{A}|$ and every $\varepsilon > 0$, there exists $\eta > 0$ such that if $\delta_{\mathcal{A}}(a, a') < \eta$, then $\delta_{\mathcal{B}}(f(a), f(a')) < \varepsilon$.*

2. *f is forward-confluent with respect to $d_{\mathcal{A}}$, $d_{\mathcal{B}}$ if and only if f is continuous and* pointwise discrete; *that is, if $a \in |\mathcal{A}|$, then there is $\varepsilon > 0$ such that if $\delta_{\mathcal{A}}(a, a') < \varepsilon$ and $f(a) = f(a')$, then $a = a'$.*

3. *f is backward-confluent with respect to $c_{\mathcal{A}}$ and $c_{\mathcal{B}}$ or, equivalently, with respect to $d_{\mathcal{A}}$ and $d_{\mathcal{B}}$, if and only if f is* open; *that is, for every $a \in |\mathcal{A}|$ and every $\varepsilon > 0$, there exists $\eta > 0$ such that if $\delta_{\mathcal{B}}(f(a), b') < \eta$, then there is $a' \in |\mathcal{A}|$ such that $\delta_{\mathcal{A}}(a, a') < \varepsilon$ and $f(a') = b'$.*

Note that item 3 does not require an additional pointwise-discreteness condition for d; the reason for this is that f is a function so that the image of any point x is discrete 'for free', as $f(\{x\})$ is always a singleton. Next, we will review some well-known constructions that yield locally bisimilar models.

4.2 Generated submodels

Given Kripke models $\mathcal{A}, \mathcal{B}$, we say that $\mathcal{A}$ is a *submodel* of $\mathcal{B}$ if $|\mathcal{A}| \subseteq |\mathcal{B}|$, $R_{\mathcal{A}} = R_{\mathcal{B}} \cap (|\mathcal{A}| \times |\mathcal{A}|)$, and $V_{\mathcal{A}}(w) = V_{\mathcal{B}}(w)$ for all $w \in |\mathcal{A}|$. It is typically not the case that $\mathcal{A}$ satisfies the same formulas as $\mathcal{B}$, unless we assume that $|\mathcal{A}|$ has some additional properties.

Definition 4.5. *If $\mathcal{B}$ is any Kripke frame or model, a set $U \subseteq |\mathcal{B}|$ is* persistent *if, whenever $w \in U$ and $w \, R_{\mathcal{B}} \, v$, it follows that $v \in U$. If $\mathcal{A}$ is a subframe (respectively, submodel) of $\mathcal{B}$, then we say that $\mathcal{A}$ is persistent if $|\mathcal{A}|$ is.*

In this case, the inclusion $\iota: |\mathcal{A}| \to |\mathcal{B}|$ is a bisimulation, and thus we obtain:

Lemma 4.6. *If $\mathcal{A}$ is a persistent submodel of $\mathcal{B}$ and $w \in |\mathcal{A}|$, then $(\mathcal{A}, w)$ is locally bisimilar to $(\mathcal{B}, w)$.*

In particular, if we are concerned with satisfiaction of $\mathcal{L}^{\mu\Diamond^*}_{\oplus\Diamond}$-formulas on a pointed model $(\mathcal{B}, w)$, it suffices to restrict our attention to the set of points accessible from w.

Definition 4.7. *Given a binary relation R, let R^* denote the transitive, reflexive closure of R.*

Then, given a Kripke frame or model $\mathcal{B}$ and $w \in |\mathcal{B}|$, we define the generated subframe (respectively, submodel) *of w to be the substructure of $\mathcal{B}$ with domain $R^*_{\mathcal{B}}(w)$.*

The following is then obvious from the definitions:

Lemma 4.8. *If $\mathcal{B}$ is a Kripke structure and $w \in |\mathcal{B}|$, then the generated substructure of w is persistent.*

Remark 4.9. *Although we will not need this in the text, persistent substructures can be generalized to other classes of convergence spaces by considering substructures with open domain (see, e.g., [21]). However, it is typically not the case that there is a least open substructure containing a given point w.*

4.3 Model amalgamation

If $\{A_i : i \in I\}$ is a family of sets, let us use $\coprod_{i\in I} A_i$ to denote its disjoint union in a standard way. We extend this notation to families $\{\mathcal{A}_i : i \in I\}$ of Kripke models by setting

$$\coprod_{i\in I} \mathcal{A}_i = (|\mathcal{A}|, R_{\mathcal{A}}, V_{\mathcal{A}}),$$

where

(i) $|\mathcal{A}| = \coprod_{i\in I} |\mathcal{A}_i|$,

(ii) $R_{\mathcal{A}} = \coprod_{i\in I} R_{\mathcal{A}_i}$, and

(iii) for $w \in |\mathcal{A}|$, $V_{\mathcal{A}}(w) = V_{\mathcal{A}_i}(w)$ if $w \in |\mathcal{A}_i|$.

It is easy to check that, for any $j \in I$, $\mathcal{A}_j$ is a persistent substructure of $\coprod_{i\in I} \mathcal{A}_i$, and thus we obtain the following from Lemma 4.2:

Lemma 4.10. *If $\{\mathcal{A}_i : i \in I\}$ is a family of models, $w \in |\mathcal{A}_j|$ and $\varphi \in \mathcal{L}^{\mu\Diamond^*}_{\oplus\Diamond}$, then $(\mathcal{A}_j, w) \models \varphi$ if and only if $(\coprod_{i\in I} \mathcal{A}_i, w) \models \varphi$.*

The tools we have presented will be instrumental throughout the text to obtain our main results. Next, we turn our attention to discussing classes of convergence spaces that will be important throughout the text.

5 Special classes of spaces

Our main succinctness results use constructions based on Kripke semantics, which we then 'lift' to other classes of spaces. Specifically, we will focus on classes of **K4** models that are confluent images of natural spaces, including Euclidean spaces. As the latter are connected and confluent maps preserve connectedness, we must work with **K4** frames that share this property.

5.1 Connectedness

Given any **K4** frame $\mathcal{A}$, there always exists some metric space $\mathcal{X}$ such that there is a surjective confluent map $f: |\mathcal{X}| \to |\mathcal{A}|$ [27]. However, if the space $\mathcal{X}$ is fixed beforehand, there is not always a guarantee that such a map exists. In particular, this is typically not the case for $\mathbb{R}^n$ for any n, due to the fact that these spaces are connected; that is, they cannot be partitioned into two disjoint open sets. More formally, $C \subseteq |\mathcal{X}|$ is *connected* if whenever $C \subseteq i(A) \cup i(\overline{A})$, it follows that either $A \cap C = \varnothing$ or $\overline{A} \cap C = \varnothing$. The space $\mathcal{X}$ is connected if $|\mathcal{X}|$ is. As observed by Shehtman, this property can be characterized using the universal modality:

Proposition 5.1 ([37]). *A convergence space $\mathcal{A}$ is connected if and only if*

$$\mathcal{A} \vDash \forall(\boxplus p \vee \boxplus \overline{p}) \to (\forall p \vee \forall \overline{p}).$$

Shehtman [36] also considered what we call *local (puncture-)connectedness,* which can be characterized with $\Box$. Say that a metric space $\mathcal{X}$ is *locally puncture-connected* if whenever $x \in |\mathcal{X}|$ and U is a neighbourhood of x, there is a neighbourhood $O \subseteq U$ of x such that $O \setminus \{x\}$ is connected. Similarly, say that a **K4** frame $\mathcal{A}$ is *locally connected* if $R_{\mathcal{A}}(a)$ is connected for all $a \in |\mathcal{A}|$.

Proposition 5.2 ([36]). *If $\mathcal{A}$ is either a locally puncture-connected metric space or a locally connected* **K4** *frame then*

$$\mathcal{A} \vDash \Box(\boxplus p \vee \boxplus \overline{p}) \to (\Box p \vee \Box \overline{p}). \tag{1}$$

It is not hard to check that $\mathbb{R}^2$ is locally puncture-connected while $\mathbb{R}$ is not. Note that there are metric spaces satisfying (1) that are not locally puncture-connected; this is studied in more detail by Lucero-Bryan [29]. In the setting of Kripke models, connectedness can be characterized by the existence of paths joining any two points:

Proposition 5.3 ([37]). *Let $\mathcal{A} = (|\mathcal{A}|, R_{\mathcal{A}})$ be a* **K4** *frame. Then, $B \subseteq |\mathcal{A}|$ is connected if and only if for all $w, v \in B$, there are $b_0 \dots b_n \in B$ such that $b_0 = w$, $b_n = v$, and for all $i < n$, either $b_i \mathrel{R_{\mathcal{A}}} b_{i+1}$ or $b_{i+1} \mathrel{R_{\mathcal{A}}} b_i$.*

Let us say that a **K4** frame $\mathcal{A}$ is *totally connected* if it is both connected and locally connected. Of particular interest to us is the class of totally connected **KD4** frames, which we denote **TC**. In view of Proposition 5.3, **TC** consists of the class of transitive, serial frames $\mathcal{A}$ such that there exists a path between any two points $w, v \in |\mathcal{A}|$ and, moreover, if $x \mathrel{R_{\mathcal{A}}} w$ and $x \mathrel{R_{\mathcal{A}}} v$, we can choose the path entirely within $R_{\mathcal{A}}(x)$.

A celebrated result of McKinsey and Tarski [30] states that any formula of $\mathcal{L}_{\oplus}$ satisfiable over an **S4** frame is satisfiable over the real line, or any other *crowded*[6] metric space $\mathcal{X}$ satisfying some natural properties. This result has since received several improvements and variations throughout the years (see e.g. [34, 5, 31, 25]). We present a powerful variant proven by Goldblatt and Hodkinson [21], which states the following.

Theorem 5.4. *Let $\mathcal{X} = (|\mathcal{X}|, d_{\mathcal{X}})$ be a crowded metric space equipped with the limit operator, and $\mathcal{A} = (|\mathcal{A}|, R_{\mathcal{A}})$ be a finite* **TC** *frame. Then, there exists a surjective, confluent map $f : |\mathcal{X}| \to |\mathcal{A}|$.*

Putting together Lemma 4.2 and Theorem 5.4, we obtain the following.

Corollary 5.5. *Let $\mathcal{X} = (|\mathcal{X}|, d_{\mathcal{X}})$ be a crowded metric space equipped with the limit operator, and $\mathcal{A} = (|\mathcal{A}|, R_{\mathcal{A}}, V_{\mathcal{A}})$ be a finite* **TC** *model. Then, there exists a map $f : |\mathcal{X}| \to |\mathcal{A}|$ and a model $\mathcal{M} = (\mathcal{X}, V_{\mathcal{M}})$ such that, for all $\varphi \in \mathcal{L}^{\mu\Diamond^*}_{\oplus\Diamond\forall}$,*

$$[\![\varphi]\!]_{\mathcal{M}} = f^{-1}[[\![\varphi]\!]_{\mathcal{A}}]. \tag{2}$$

Proof. The map f is the surjective, confluent map provided by Theorem 5.4, and the valuation $V_{\mathcal{X}}$ is defined by $p \in V_{\mathcal{X}}(x)$ if and only if $p \in V_{\mathcal{A}}(f(x))$ for $p \in P$ and $x \in |\mathcal{X}|$. That (2) holds follows from Lemma 4.2. □

5.2 Scattered spaces

In *provability logic* [7], a seemingly unrelated application of modal logic, $\Box\varphi$ is interpreted as 'φ is a theorem of (say) Peano arithmetic'. It follows from a result of Solovay [38] that, surprisingly, the valid formulas under this interpretation are exactly the valid formulas over the class of scattered limit spaces. This non-trivial link

[6] A metric space is *crowded* if $d_{\mathcal{X}}(|\mathcal{X}|) = |\mathcal{X}|$; i.e., if $\mathcal{X}$ contains no isolated points.

between proof theory and spatial reasoning allows for an additional and unexpected application of the logics we are considering. For this, let us define scattered spaces in the context of convergence spaces.

Definition 5.6. *A convergence space $\mathcal{A}$ is* scattered *if, for every $X \subseteq |\mathcal{A}|$, if $X \subseteq \rho_{\mathcal{A}}(X)$, then $X = \varnothing$.*

In other words, if $X \neq \varnothing$, then there is $a \in X \setminus \rho_{\mathcal{A}}(X)$; such a point is an *isolated point* of X. This property is characterized by the well-known Löb axiom:

Proposition 5.7 ([4]). *A convergence space $\mathcal{A}$ is scattered if and only if*

$$\Box(\Box p \to p) \to \Box p$$

is valid on $\mathcal{A}$.

It is not difficult to produce examples of scattered spaces; the most standard are based either on ordinals or on Kripke frames. Let us begin with the latter.

Lemma 5.8. *Let $\mathcal{A} = (|\mathcal{A}|, R_{\mathcal{A}})$ be any* **K4** *frame. Then, $\mathcal{A}$ is a scattered space if and only if $R_{\mathcal{A}}$ is* converse-well-founded; *that is, there are no infinite sequences $a_0 \; R_{\mathcal{A}} \; a_1 \; R_{\mathcal{A}} \; a_2 \; R_{\mathcal{A}} \ldots$.*

In particular, if $\mathcal{A}$ is a finite **K4** frame, then $\mathcal{A}$ is scattered as a convergence space if and only if $R_{\mathcal{A}}$ is irreflexive. The class of frames with a transitive, converse well-founded relation is named **GL** after Gödel and Löb, whose contributions led to the development of provability logic. Note that the tangled operator is not very interesting in this setting.

Proposition 5.9. *Let $\varphi_1, \ldots, \varphi_n \in \mathcal{L}^{\mu\Diamond^*}_{\oplus\Diamond\forall}$. Then, $\Diamond^*\{\varphi_1, \ldots, \varphi_n\} \equiv \bot$ over the class of scattered spaces.*

Proof. Assume that $(\mathcal{A}, V)$ is a scattered limit model, and consider a set S such that, for $i \leq n$, $S \subseteq \rho_{\mathcal{A}}(S \cap [\![\varphi_i]\!]_V)$. But, this implies that $S \subseteq \rho_{\mathcal{A}}(S)$, which, since $\mathcal{A}$ is scattered, means that $S = \varnothing$. Since $[\![\Diamond^*\{\varphi_1, \ldots, \varphi_n\}]\!]_V$ is the union of all such S, we conclude that $[\![\Diamond^*\{\varphi_1, \ldots, \varphi_n\}]\!]_V = \varnothing$. □

From this we immediately obtain the following.

Corollary 5.10. *There exists a function $t^{\bot}_{\Diamond^*} : \mathcal{L}^{\Diamond^*}_{\Diamond\forall} \to \mathcal{L}_{\Diamond\forall}$ such that $\varphi \equiv t^{\bot}_{\Diamond^*}(\varphi)$ is valid over the class of scattered spaces and $\left|t^{\bot}_{\Diamond^*}(\varphi)\right| \leq |\varphi|$ for all $\varphi \in \mathcal{L}^{\Diamond^*}_{\oplus\forall}$.*

Theorem 5.4 has an analogue for a family of 'nice' scattered spaces. Below and throughout the text, we use the standard set-theoretic convention that an ordinal is equivalent to the set of ordinals below it, i.e. $\zeta \in \xi$ if and only if $\zeta < \xi$.

Definition 5.11. *Given an ordinal* Λ, *define* $d: 2^{\Lambda} \to 2^{\Lambda}$ *by letting* $\xi \in d(X)$ *if and only if* $X \cap \xi$ *is unbounded in* ξ.

Recall that addition, multiplication and exponentiation are naturally defined on the ordinal numbers (see, e.g., [24]) and that ω defines the least infinite ordinal. The following result can be traced back to Abashidze [1] and Blass [6], and is proven in a more general form by Aguilera and Fernández-Duque [3].

Theorem 5.12. *If* $\mathcal{A}$ *is any finite* **GL** *frame, then there exists an ordinal* $\Lambda < \omega^{\omega}$ *and a surjective map* $f: \Lambda \to |\mathcal{A}|$ *that is confluent with respect to the limit operator on* Λ.

As before, this readily gives us the following corollary:

Corollary 5.13. *Given a finite* **GL**-*model* $\mathcal{A} = (|\mathcal{A}|, R_{\mathcal{A}}, V_{\mathcal{A}})$, *there exists an ordinal* $\Lambda < \omega^{\omega}$, *a surjective map* $f: \Lambda \to |\mathcal{A}|$, *and a model* $\mathcal{M} = (\Lambda, V_{\mathcal{M}})$ *such that, for all* $\varphi \in \mathcal{L}^{\mu\Diamond^*}_{\oplus\Diamond\forall}$,

$$[\![\varphi]\!]_{\mathcal{M}} = f^{-1}[[\![\varphi]\!]_{\mathcal{A}}].$$

Now that we have settled the classes of structures we are interested in, we discuss the techniques that we will use to establish our main succinctness results.

6 Model equivalence games

The limit-point, or set-derivative, operator $\Diamond$ is strictly more expressive than the closure operator $\oplus$ [4]. Nevertheless, if we consider a formula such as $\varphi = \underbrace{\oplus\oplus\ldots\oplus}_{n}\top$, we observe that its translation $t^{\Diamond}_{\oplus}(\varphi)$ into $\mathcal{L}_{\Diamond}$ is exponential. Of course in this case $\varphi \equiv \top$, but as we will see in Section 7, this exponential blow-up is inevitable for other choices of φ. To be precise, we wish to show that there is no translation $t: \mathcal{L}_{\oplus} \to \mathcal{L}_{\Diamond}$ for which there exists a sub-exponential function $f: \mathbb{N} \to \mathbb{N}$ such that $t(\varphi) \equiv \varphi$ over the class of convergence spaces and $|t(\varphi)| \leq f(|\varphi|)$. In view of Theorem 5.10, to show that $\varphi \not\equiv \psi$ over the class of convergence spaces (or even metric spaces), it suffices to exhibit a model $\mathcal{A} \in \mathbf{K4}$ and $a \in |\mathcal{A}|$ such that $(\mathcal{A}, a) \models \varphi$ but $(\mathcal{A}, a) \not\models \psi$, or vice-versa. We will prove that such $\mathcal{A}$ exists whenever ψ is small by using *model equivalence games,* which are based on sets of pointed models.

We will use $\boldsymbol{a}, \boldsymbol{b}, \ldots$ to denote pointed models. As was the case for non-pointed structures, for a class of pointed models $\mathbf{A}$ and a formula φ, we write $\mathbf{A} \models \varphi$ when $\boldsymbol{a} \models \varphi$ for all $\boldsymbol{a} \in \mathbf{A}$, i.e., φ is true in any pointed model in $\mathbf{A}$, and say that the formulas φ and ψ are equivalent on a class of pointed models $\mathbf{A}$ when $\boldsymbol{a} \models \varphi$ if and only if $\boldsymbol{a} \models \psi$ for all pointed models $\boldsymbol{a} \in \mathbf{A}$. We can also define an accessibility relation between pointed models.

Definition 6.1. *For a pointed model $\boldsymbol{a} = (\mathcal{A}, a)$, we denote by $\Box\boldsymbol{a}$ the set $\{(\mathcal{A}, b) : a \, R_{\mathcal{A}} \, b\}$, i.e., the set of all pointed models that are successors of the pointed model $\boldsymbol{a}$ along the relation $R_{\mathcal{A}}$.*

The game described below is essentially the formula-size game from Adler and Immerman [2] but reformulated slightly to fit our present purposes. The general idea is that we have two competing players, Hercules and the Hydra. Given a formula φ and a class of pointed models $\mathbf{M}$, Hercules is trying to show that there is a "small" formula ψ in the language $\mathcal{L}_{\Diamond}$ that is equivalent to φ on $\mathbf{M}$, whereas the Hydra is trying to show that any such ψ is "big". Of course, what "small" and "big" mean depends on the context at hand. The players move by adding and labelling nodes on a game-tree, T. Although our use of trees is fairly standard, they play a prominent role throughout the text, so let us give some basic definitions before setting up the game.

Definition 6.2. *For our purposes, a* tree *is a pair $(T, <)$, where T is any set and $<$ a strict partial order such that, if $\eta \in T$, then $\{\zeta \in T : \zeta < \eta\}$ is finite and linearly ordered, and T has a least element called its* root*. We will sometimes notationally identify $(T, <)$ with T, and write $\preccurlyeq$ for the reflexive closure of $<$.*

Maximal elements of T are leaves*. If $\zeta, \eta \in T$, we say that η is a* daughter *of ζ if $\zeta < \eta$ and there is no ξ such that $\zeta < \xi < \eta$. A* path (of length m) *on T is a sequence $\vec{\eta} = (\eta_i)_{i \leq m}$ such that η_{i+1} is a daughter of η_i whenever $i < m$.*

Next, Definition 6.3 gives the precise moves that Hercules and the Hydra may play in the game.

Definition 6.3. *Let $\mathbf{M}$ be a class of pointed models and φ be a formula. The (φ, $\mathbf{M}$)* model equivalence game *(($\varphi, \mathbf{M}$)-MEG) is played by two players, Hercules and the Hydra, according to the following instructions.*

SETTING UP THE PLAYING FIELD. *The Hydra initiates the game by choosing two classes of pointed models $\mathbf{A}, \mathbf{B} \subseteq \mathbf{M}$ such that $\mathbf{A} \models \varphi$ and $\mathbf{B} \models \neg\varphi$.*

After that, the players continue the $(\varphi, \mathbf{M})$-MEG *on the pair* $(\mathbf{A}, \mathbf{B})$ *by constructing a finite game-tree* T*, in such a way that each node* $\eta \in T$ *is labelled with a pair* $(\mathfrak{L}(\eta), \mathfrak{R}(\eta))$ *of classes of pointed models and one symbol that is either a literal or one from the set* $\{\vee, \wedge, \Box, \Diamond, \exists, \forall\}$. *We will usually write* $\mathfrak{L}(\eta) \circ \mathfrak{R}(\eta)$ *instead of* $(\mathfrak{L}(\eta), \mathfrak{R}(\eta))$*, where the symbol '*$\circ$*' is meant to be suggestive of a node (not to be confused with composition). The pointed models in* $\mathfrak{L}(\eta)$ *are called* the models on the left, *and those in* $\mathfrak{R}(\eta)$ *are called* the models on the right.

Any leaf η *can be declared either a* head *or a* stub. *Once* η *has been declared a stub, no further moves can be played on it. The construction of the game-tree begins with a root labeled by* $\mathbf{A} \circ \mathbf{B}$ *that is declared a head.*

Afterwards, the game continues as long as there is at least one head. In each turn, Hercules goes first by choosing a head η*, labeled by* $\mathbf{L} \circ \mathbf{R} = \mathfrak{L}(\eta) \circ \mathfrak{R}(\eta)$. *He then plays one of the following moves.*

LITERAL-MOVE. *Hercules chooses a literal* ι *such that* $\mathbf{L} \models \iota$ *and* $\mathbf{R} \not\models \iota$. *The node* η *is declared a stub and labelled with the symbol* ι.

$\vee$-MOVE. *Hercules labels* η *with the symbol* $\vee$ *and chooses two sets* $\mathbf{L}_1, \mathbf{L}_2 \subseteq \mathbf{L}$ *such that* $\mathbf{L} = \mathbf{L}_1 \cup \mathbf{L}_2$. *Two new heads, labeled by* $\mathbf{L}_1 \circ \mathbf{R}$ *and* $\mathbf{L}_2 \circ \mathbf{R}$*, are added to the tree as daughters of* η.

$\wedge$-MOVE. *Hercules labels* η *with the symbol* $\wedge$ *and chooses two sets* $\mathbf{R}_1, \mathbf{R}_2 \subseteq \mathbf{R}$ *such that* $\mathbf{R} = \mathbf{R}_1 \cup \mathbf{R}_2$. *Two new heads, labeled by* $\mathbf{L} \circ \mathbf{R}_1$ *and* $\mathbf{L} \circ \mathbf{R}_2$*, are added to the tree as daughters of* η.

$\Diamond$-MOVE. *Hercules labels* η *with the symbol* $\Diamond$ *and, for each pointed model* $\boldsymbol{l} \in \mathbf{L}$*, he chooses a pointed model from* $\Box \boldsymbol{l}$ *(if for some* $\boldsymbol{l} \in \mathbf{L}$ *we have* $\Box \boldsymbol{l} = \varnothing$*, Hercules cannot play this move). All these new pointed models are collected in the set* $\mathbf{L}_1$. *For each pointed model* $\boldsymbol{r} \in \mathbf{R}$*, the Hydra replies by picking a subset of* $\Box \boldsymbol{r}$.[7] *All the pointed models chosen by the Hydra are collected in the class* $\mathbf{R}_1$. *A new head labeled by* $\mathbf{L}_1 \circ \mathbf{R}_1$ *is added as a daughter to* η.

$\Box$-MOVE. *Hercules labels* η *with the symbol* $\Box$ *and, for each pointed model* $\boldsymbol{r} \in \mathbf{R}$*, he chooses a pointed model from* $\Box \boldsymbol{r}$ *(as before, if for some* $\boldsymbol{r} \in \mathbf{R}$ *we have that* $\Box \boldsymbol{r} = \varnothing$*, then Hercules cannot play this move). All these new pointed models are collected in the set* $\mathbf{R}_1$. *The Hydra replies by constructing a class of models* $\mathbf{L}_1$ *as follows. For*

[7]In particular, if $\Box \boldsymbol{r} = \varnothing$ for some $\boldsymbol{r}$, the Hydra does not add anything to $\mathbf{R}_1$ for the pointed model $\boldsymbol{r}$.

each $\boldsymbol{l} \in \mathbf{L}$*, she picks a subset of* $\Box \boldsymbol{l}$ *and collects all these pointed models in the set* $\mathbf{L}_1$*. A head labeled by* $\mathbf{L}_1 \circ \mathbf{R}_1$ *is added as a daughter to* η*.*

The $(\varphi, \mathbf{M})$-MEG *game concludes when there are no heads. Hercules has a winning strategy in* n *moves in the* $(\varphi, \mathbf{M})$-MEG *iff no matter how the Hydra plays, the resulting game tree has* n *nodes and there are no heads; note that we do not count the move performed by the Hydra when setting up the playing field.*

We remark that $\wedge$- and $\Box$-moves are symmetric to $\vee$- moves and $\Diamond$-moves, respectively, except that the roles of the left- and the right-hand sides are reversed. The relation between the $(\varphi, \mathbf{M})$-MEG and formula-size is given by the following result. The essential features of the proof of the next theorem can be found in any one of [16, 17, 23].

Theorem 6.4. *Hercules has a winning strategy in* n *moves in the* $(\varphi, \mathbf{M})$-MEG *iff there is a* $\mathcal{L}_\Diamond$*-formula* ψ *with* $|\psi| \le n$ *that is equivalent to* φ *on* $\mathbf{M}$*.*

Intuitively, a winning strategy for Hercules in the $(\varphi, \mathbf{M})$-MEG is given by the syntax tree T_ψ of any $\mathcal{L}_\Diamond$-formula ψ that is equivalent to φ on $\mathbf{M}$ (note that φ is not necessarily a modal formula: for example, it can be a first- or second-order formula, or a formula from a different modal language). Since Hercules is trying to prove that there *exists* a small formula, i.e., the number of nodes in its syntax tree is small, while the Hydra is trying to show that any such formula is "big", if both Hercules and the Hydra play optimally and Hercules has a winning strategy, then the resulting game tree T is the syntax tree T_ψ of a minimal modal formula ψ that is equivalent to φ on $\mathbf{C}$. In particular, if $\mathbf{L} \circ \mathbf{R}$ is the label of the root, we have that ψ must be true in every element of $\mathbf{L}$ and false on every element of $\mathbf{R}$, which would be impossible if there were two bisimilar pointed models $\boldsymbol{l} \in \mathbf{L}$ and $\boldsymbol{r} \in \mathbf{R}$. More generally, the subformula θ of ψ corresponding to any node η is true on all pointed models of $\mathfrak{L}(\eta)$ and false on all pointed models of $\mathfrak{R}(\eta)$. It follows that Hercules loses if any node has bisimilar models on the left and right, provided the Hydra plays well.

As for what it means to 'play well', note that the Hydra has no incentive to pick less pointed models in her turns, so it suffices to assume that she plays *greedily:*

Definition 6.5. *We say that the Hydra* plays greedily *if:*

(a) whenever Hercules makes a $\Diamond$*-move on a head* η*, the Hydra replies by choosing all of* $\Box \boldsymbol{b}$ *for each* $\boldsymbol{b} \in \mathfrak{R}(\eta)$*, and*

(b) whenever Hercules makes a $\Box$*-move on a head* η*, the Hydra replies by choosing all of* $\Box \boldsymbol{a}$ *for each* $\boldsymbol{a} \in \mathfrak{L}(\eta)$*.*

If the Hydra plays greedily, Hercules must avoid having bisimilar models on each side:

Lemma 6.6. *If the Hydra plays greedily, no closed game tree contains a node η such that there are $\boldsymbol{l} \in \mathfrak{L}(\eta)$ and $\boldsymbol{r} \in \mathfrak{R}(\eta)$ that are locally bisimilar.*

Proof. We prove by induction on the number of rounds in the game that, once a node η_0 such that there are bisimilar $\boldsymbol{l} \in \mathfrak{L}(\eta_0)$ and $\boldsymbol{r} \in \mathfrak{R}(\eta_0)$ is introduced, there will always be a head η with bisimilar pointed models on each side. The base case, where η_0 is first introduced, is trivial, as new nodes are always declared to be heads.

For the inductive step, assume that there are a head η, $\boldsymbol{l} = (\mathcal{L}, l) \in \mathfrak{L}(\eta)$, $\boldsymbol{r} = (\mathcal{R}, r) \in \mathfrak{R}(\eta)$, and a bisimulation $\chi \subseteq |\mathcal{L}| \times |\mathcal{R}|$ with $l \ \chi \ r$. We may also assume that Hercules plays on η, since otherwise η remains on the tree as a head.

Hercules cannot play a literal move on η since l and r agree on all atoms. If Hercules plays an $\vee$-move, he chooses $\mathbf{L}_1, \mathbf{L}_2$ so that $\mathbf{L}_1 \cup \mathbf{L}_2 = \mathfrak{L}(\eta)$ and creates two new nodes η_1 and η_2 labeled by $\mathbf{L}_1 \circ \mathfrak{R}(\eta)$ and $\mathbf{L}_2 \circ \mathfrak{R}(\eta)$, respectively. If $\boldsymbol{l} \in \mathbf{L}_1$, then we observe that $(\boldsymbol{l}, \boldsymbol{r})$ is a pair of bisimilar pointed models that still appears in $\mathbf{L}_1 \circ \mathfrak{R}(\eta)$. If not, then $\boldsymbol{l} \in \mathbf{L}_2$, and the pair appears on $\mathbf{L}_2 \circ \mathfrak{R}(\eta)$. The case for an $\wedge$-move is symmetric.

If Hercules plays a $\Diamond$-move, then for the pointed model $(\mathcal{L}, l') \in \Box \boldsymbol{l}$ that Hercules chooses, by forward confluence and the assumption that the Hydra plays greedily, the Hydra will choose at least one pointed model $(\mathcal{R}, r') \in \Box \boldsymbol{r}$ such that $l' \ \chi \ r'$. Similarly, if Hercules plays a $\Box$-move, then for the pointed model $(\mathcal{R}, r') \in \Box \boldsymbol{r}$ that Hercules chooses, using backwards confluence, the Hydra replies by choosing at least one $(\mathcal{L}, l') \in \Box \boldsymbol{l}$ such that $l' \ \chi \ r'$. It follows that, no matter how Hercules plays, the following state in the game will also contain two bisimilar pointed models, and hence the game-tree will never be closed. □

Example 6.7. *A closed game tree for a model equivalence game is shown in Figure 2. Pointed models occurring along the nodes of the tree are pairs consisting of the relevant model $\mathcal{A}_1$, $\mathcal{A}_2$ or $\mathcal{B}$ and the nodes marked by $\triangleright$. The relations between the points in the respective Kripke frame are denoted by the arrows, i.e., if $\mathcal{F} \in \{\mathcal{A}_1, \mathcal{A}_2, \mathcal{B}\}$, then, for $w, v \in |\mathcal{F}|$, we have $w \ R_{\mathcal{F}} \ v$ if and only if there is an arrow coming out of w and pointing to v. We have only one proposition p which is true only on the black points. Note that, if we disregard the Kripke models, the game tree is actually the syntax tree of the formula $\Box p \vee \Diamond\Diamond p$. It is easily seen that, for any node η in the tree, the sub-formula of $\Box p \vee \Diamond\Diamond p$ starting at η is true in all the pointed models on the left of η and false in all the pointed models on the right. It is worth pointing out that Hercules could have also won if he had played according to the formula $\Box p \vee \Diamond(\overline{p} \wedge \Diamond p)$.*

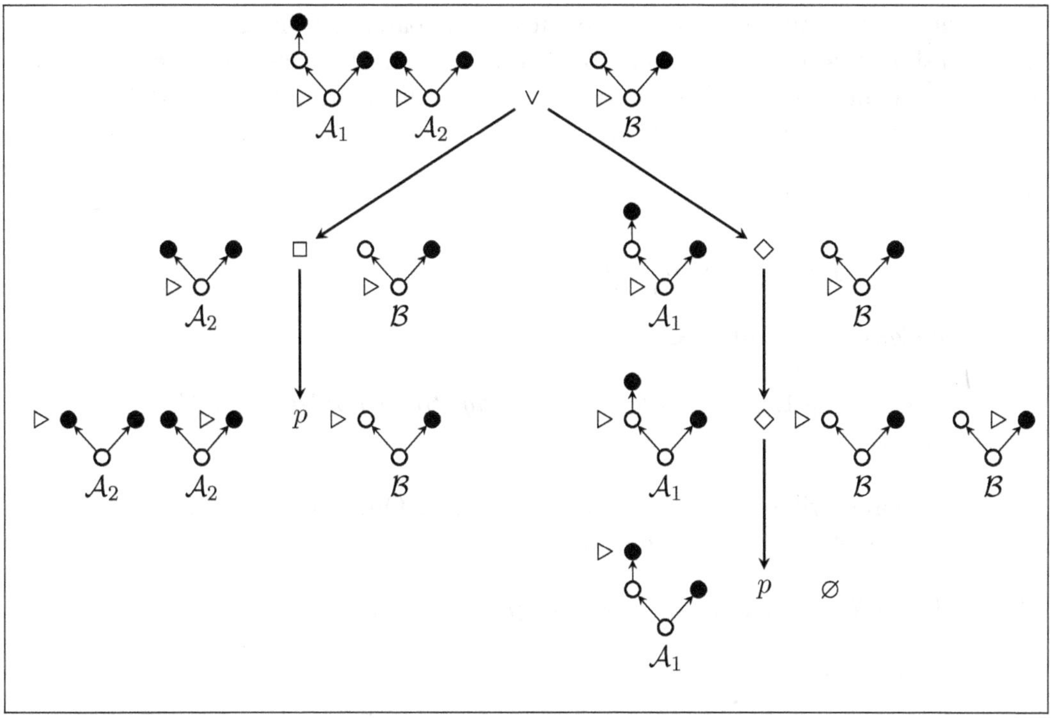

Figure 2: A closed game tree.

7 Exponential succinctness of closure over derivative

Now we may use the model equivalence games we have presented to show that the closure operation is exponentially more succinct than the set-derivative operator. Our proof will be based on the following infinite sequence of formulas.

Definition 7.1. *For every $n \geq 1$, let the formulas φ_n be defined recursively by*

(i) $\varphi_1 = \diamondplus p_1$, and *(ii) $\varphi_{n+1} = \diamondplus(p_{n+1} \wedge \varphi_n)$.*

Then, define $\psi_n = t^{\Diamond}_{\diamondplus}(\varphi_n)$.

Recall that $t^{\Diamond}_{\diamondplus}$ was defined in Section 3.1, and that ϕ_n is equivalent to ψ_n for every n. Before we proceed, let us give some bounds on the size of the formulas we have defined, which can be proven by an easy induction.

Lemma 7.2. *For all $n \in \mathbb{N}$, $|\varphi_n| \leq 3n$ and $|\psi_n| \geq 2^n$.*

Thus there is an exponential blow-up when passing from φ_n to ψ_n. We are going to show that on any class of models that contains the class of finite **GL** or **TC** models, we cannot find an essentially shorter formula than ψ_n in the modal language $\mathcal{L}_\Diamond$ that is equivalent to φ_n. This result will be a consequence of the following.

Theorem 7.3. *Let* **C** *be either:*

(a) the class of all finite **GL** *frames, or*

(b) the class of all finite **TC** *frames.*

Then, for every $n \geq 1$, Hercules has no winning strategy of less than 2^n moves in the $(\varphi_n, \mathbf{C})$-MEG.

This theorem will be proven later in this section. Once we do, we will immediately obtain a series of succinctness results:

Proposition 7.4. *Let* **C** *be a class of convergence spaces containing either*

1. *all finite* **GL** *frames,*

2. *all finite* **TC** *frames,*

3. *all ordinals $\Lambda < \omega^\omega$, or*

4. *any crowded metric space $\mathcal{X}$.*

Then, for all $n \geq 1$, whenever $\psi \in \mathcal{L}_\Diamond$ is equivalent to φ_n over **C***, it follows that $|\psi| \geq 2^{\frac{|\varphi|}{3}}$.*

Proof. In the first two cases, the claim follows immediately from Theorems 7.3 and 6.4.

Now, suppose that $|\psi| \leq 2^{\frac{|\varphi_n|}{3}}$. Then, by the first claim, there is a finite, pointed **GL** model $(\mathcal{K}, w)$ such that $(\mathcal{K}, w) \not\models \psi \leftrightarrow \varphi_n$. By Corollary 5.13, there are a model $\mathcal{M}$ based on some $\Lambda < \omega^\omega$ and a surjective map $f\colon \Lambda \to |\mathcal{K}|$ such that $[\![\theta]\!]_\mathcal{M} = f^{-1}[[\![\theta]\!]_\mathcal{K}]$ for all Θ. In particular, for $\xi \in f^{-1}(w)$ we have that $(\mathcal{M}, \xi) \not\models \psi \leftrightarrow \varphi_n$, and thus ψ is not equivalent to φ_n on ω^ω.

If **C** contains a crowded metric space $\mathcal{X}$, we reason analogously, but instead choose $\mathcal{K}$ to be a **TC**-model and use Corollary 5.5 to produce the required function f. □

Proposition 7.4 will be progressively improved throughout the text until culminating in Theorem 8.6. In order to prove Theorem 7.3 when $\mathbf{C}$ is the class of finite **GL**-models, we are going to define, for every $n \geq 1$, a pair of sets of pointed **GL**-models $\mathbf{A}^n$, $\mathbf{B}^n$ such that $\mathbf{A}^n \models \varphi_n$, whereas $\mathbf{B}^n \models \neg\varphi_n$. The Hydra is going to pick $\mathbf{A}^n$ and $\mathbf{B}^n$ when setting up the playing field for the $(\varphi_n, \mathbf{GL})$-MEG. After that, we show that Hercules has no winning strategy of less than 2^n moves.

7.1 The sets of models $\mathbf{A}^n$ and $\mathbf{B}^n$

Each model in $\mathbf{A}^n \cup \mathbf{B}^n$ is based on a finite, transitive, irreflexive tree (i.e., a finite, tree-like **GL** model), as illustrated in Figures 3 and 4. The 'critical' part of each model lies in its right-most branch as shown in the figures. To formalize this, let us begin with a few basic definitions.

Definition 7.5. *A model $\mathcal{A}$ is* rooted *if there is a unique $w_0 \in |\mathcal{A}|$ with $|\mathcal{A}| = \{w_0\} \cup R_{\mathcal{A}}(w_0)$, and* tree-like *if $(|\mathcal{A}|, R_{\mathcal{A}})$ is a tree. A* model with successors *is a model $\mathcal{A}$ equipped with a partial function $S_{\mathcal{A}}: |\mathcal{A}| \to |\mathcal{A}|$ such that $S_{\mathcal{A}}(a)$ is always a daughter of a.*

If $\mathcal{A}$ is a rooted model with successors, the critical branch *of $\mathcal{A}$ is the maximal path $\vec{w} = (w_i)_{i \leq m}$ such that w_0 is the root of $\mathcal{A}$ and $w_{i+1} = S_{\mathcal{A}}(w_i)$ for all $i < m$; we say that m is the* critical height *of $\mathcal{A}$.*

We denote the generated submodel of $S_{\mathcal{A}}(w_0)$ by $S[\mathcal{A}]$. If $w \in |\mathcal{A}|$ and $S_{\mathcal{A}}(w)$ is defined, we define $S[(\mathcal{A}, w)] = (\mathcal{A}, S_{\mathcal{A}}(w))$; note that in this case, we do not restrict the domain.

Observe that the notation $S[\cdot]$ has a different meaning depending on whether the argument is a model or a pointed model; these conventions will be helpful in the rest of the text. If $\mathcal{A}$ is a non-pointed model, then $S[\mathcal{A}]$ is a smaller, non-pointed model; however, if $\boldsymbol{a}$ is a *pointed* model, then $S[\boldsymbol{a}]$ is identical to $\boldsymbol{a}$ except for the evaluation point. The partial function $S_{\mathcal{A}}$ will not be used in the semantics, but it will help us to describe Hercules' strategy. Let us begin by defining recursively the two sets $\mathbf{A}^n$ and $\mathbf{B}^n$, each containing 2^n pointed models with successors. The formal definition of $\mathbf{A}^n$ and $\mathbf{B}^n$ is as follows.

Definition 7.6. *For $n \geq 0$, the **GL**-models in the sets $\mathbf{A}^{n+1} = \{\mathcal{A}_i^{n+1} : i \leq 2^n\}$ and $\mathbf{B}^{n+1} = \{\mathcal{B}_i^{n+1} : i \leq 2^{n+1}\}$ are defined recursively according to the following cases. When defined, we will denote the roots of $\mathcal{A}_j^m$ and $\mathcal{B}_j^m$ by a_j^m and b_j^m, respectively, and the pointed models $(\mathcal{A}_j^m, a_j^m)$, $(\mathcal{B}_j^m, b_j^m)$ by $\boldsymbol{a}_j^m$ and $\boldsymbol{b}_j^m$.*

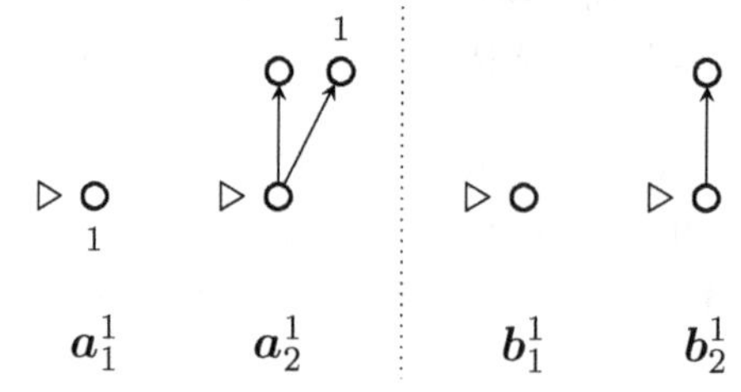

Figure 3: The pointed models in the sets $\mathbf{A}^1$ and $\mathbf{B}^1$. The numbers appearing next to each point w are the indices of the propositional variables true in w. Intuitively, $\mathcal{B}_2^1$ is obtained by "erasing" the right branch of $\mathcal{A}_2^1$.

CASE $i \leq 2^n$. *If $n = 0$, then $\mathcal{A}_1^1$ is a single point a_1^1 with $V_{\mathcal{A}_1^1}(a_1^1) = \{p_1\}$, and $\mathcal{B}_1^1$ is a single point b_1^1 with $V_{\mathcal{B}_1^1}(b_1^1) = \varnothing$.*

If $n > 0$, assume inductively that $\mathbf{A}^n, \mathbf{B}^n$ have been defined. We define $\mathcal{A}_i^{n+1}$ to be a copy of $\mathcal{A}_i^n$, except that the new propositional symbol p_{n+1} is true in the root. Similarly, $\mathcal{B}_i^{n+1}$ is a copy of $\mathcal{B}_i^n$, except that p_{n+1} is true in the root.

CASE $i > 2^n$. *Let $j = i - 2^n$. Set*

$$\mathcal{X} = \Big(\coprod_{k=2}^{2^n} S[\mathcal{A}_k^{n+1}]\Big) \amalg \mathcal{B}_j^{n+1},$$

and then construct $\mathcal{B}_{2^n+j}^{n+1}$ by adding a (fresh) irreflexive root $b_{2^n+j}^{n+1}$ to $\mathcal{X}$ which sees all elements of $|\mathcal{X}|$ and satisfies no atoms. We set

$$S_{\mathcal{B}_{2^n+j}^{n+1}} = S_{\mathcal{B}_j^{n+1}} \cup \{(b_{2^n+j}^{n+1}, b_j^{n+1})\}.$$

The models $\mathcal{A}_{2^n+j}^{n+1}$ are constructed similarly, except that we take

$$\mathcal{Y} = \Big(\coprod_{k=2}^{2^n} S[\mathcal{A}_k^{n+1}]\Big) \amalg \mathcal{B}_j^{n+1} \amalg \mathcal{A}_j^{n+1},$$

and add an irreflexive root which sees all elements of $\mathcal{Y}$ and satisfies no atoms. Finally, we set

$$S_{\mathcal{A}_{2^n+j}^{n+1}} = S_{\mathcal{A}_j^{n+1}} \cup \{(a_{2^n+j}^{n+1}, a_j^{n+1})\}.$$

Example 7.7. *Figure 3 shows $\mathbf{A}^1$ and $\mathbf{B}^1$. We are using the conventions established in Example 6.7 that each pointed model consists of the relevant model and the point*

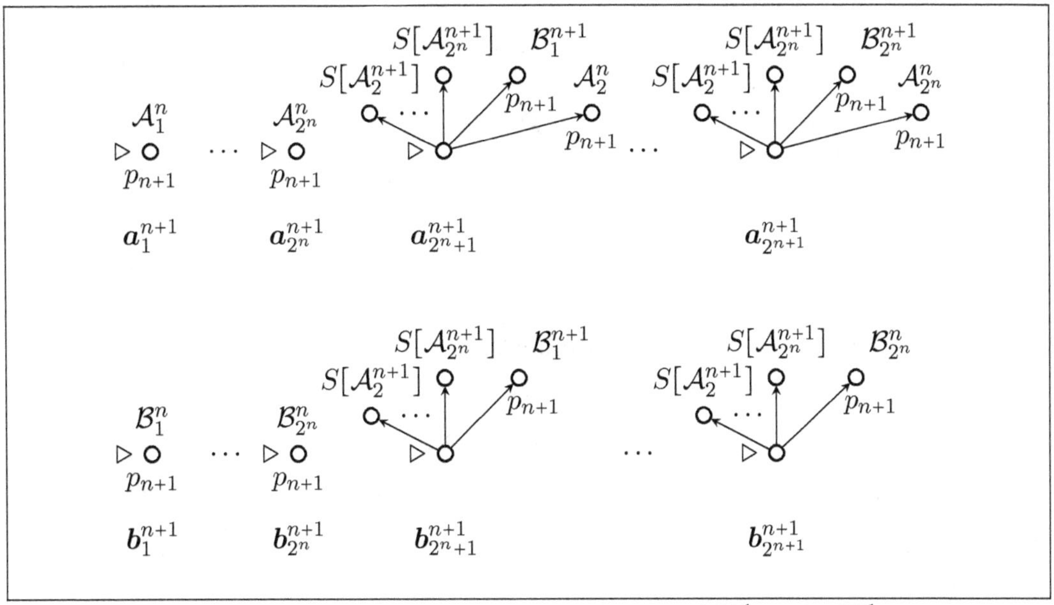

Figure 4: The pointed models in the sets $\mathbf{A}^{n+1}$ and $\mathbf{B}^{n+1}$.

designated with $\triangleright$. *The indices of the propositional letters that are true on a point are shown next to it. In any frame* $\mathcal{F}$ *displayed and any* $w \in |\mathcal{F}|$, $S_{\mathcal{F}}(w)$ *is the right-most daughter of* w *(when it exists). Note that* $\mathcal{A}_2^1$ *and* $\mathcal{B}_2^1$ *are defined according to the inductive clause, where in the latter* $\mathcal{X}$ *is just* $\mathcal{B}_1^1$ *(as the rest of the disjoint union has empty range) and* $\mathcal{Y}$ *is* $\mathcal{B}_1^1 \sqcup \mathcal{A}_1^1$.

Next, $\mathbf{A}^2$ *and* $\mathbf{B}^2$ *are shown in Figure 5, and are obtained as follows.* $\mathcal{A}_1^2, \mathcal{A}_2^2$ *are copies of* $\mathcal{A}_1^1, \mathcal{A}_2^1$, *but with* p_2 *made true at the root, and* $\mathcal{B}_1^2, \mathcal{B}_2^2$ *are defined similarly from* $\mathcal{B}_1^1, \mathcal{B}_2^1$. *In this case, we just have that* $\coprod_{k=2}^{2^1} S[\mathcal{A}_k^2] = S[\mathcal{A}_1^2]$, *so that for example* $\mathcal{B}_3^2$ *is obtained by taking* $\mathcal{X} = S[\mathcal{A}_1^2] \sqcup \mathcal{B}_3^2$ *and* $\mathcal{A}_3^2$ *is obtained by taking* $\mathcal{Y} = S[\mathcal{A}_1^2] \sqcup \mathcal{B}_3^2 \sqcup \mathcal{A}_3^2$. *Note that we do not take* $k = 1$ *in the disjoint union, as* $S[\mathcal{A}_1^1]$ *is not defined.*

Finally, let us show how to obtain the models in $\mathbf{A}^3$ *and* $\mathbf{B}^3$ *with the help of the models in* $\mathbf{A}^2$ *and* $\mathbf{B}^2$ *(see Figure 6). Note that the relations denoted by the arrows are actually transitive but the remaining arrows are not shown in order to avoid cluttering. As before, the first four pointed models in* $\mathbf{A}^3$ *and* $\mathbf{B}^3$ *are obtained from the four models in* $\mathbf{A}^2$ *and* $\mathbf{B}^2$, *respectively, by simply making the new proposition* p_3 *true in their roots. To construct the next four models in* $\mathbf{A}^3$ *and* $\mathbf{B}^3$, *observe that* $\coprod_{k=2}^{2^2} S[\mathcal{A}_k^3] = S[\mathcal{A}_2^3] \sqcup S[\mathcal{A}_3^3] \sqcup S[\mathcal{A}_4^3]$. *Then,* $\mathcal{B}_{4+j}^3$ *is defined by adding a root to* $\mathcal{X} = S[\mathcal{A}_2^3] \sqcup S[\mathcal{A}_3^3] \sqcup S[\mathcal{A}_4^3] \sqcup \mathcal{B}_j^3$, *and* $\mathcal{A}_{4+j}^3$ *is obtained by adding* $\mathcal{A}_j^3$ *to* $\mathcal{B}_j^3$.

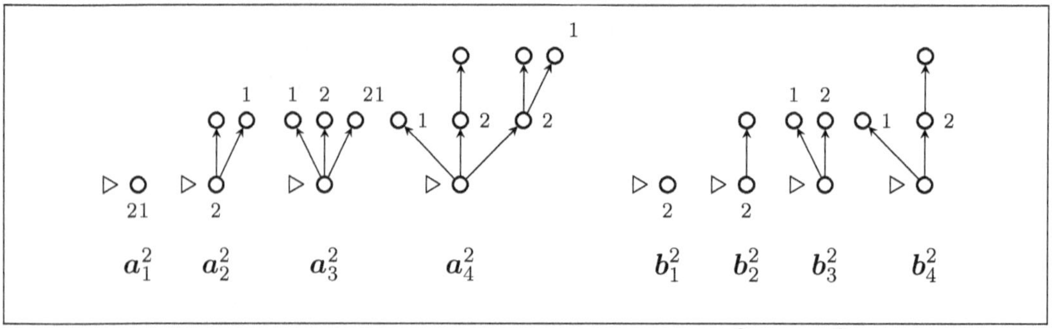

Figure 5: The pointed models in the sets $\mathbf{A}^2$ and $\mathbf{B}^2$.

Finally, we remark that $S[\mathcal{A}_1^3] \cong \mathcal{A}_1^1$, $S[\mathcal{A}_2^3] \cong \mathcal{A}_1^2$ *and* $S[\mathcal{A}_3^3] \cong \mathcal{A}_2^2$.

In fact, the latter observation is only a special case of a general pattern.

Lemma 7.8. *Let* $i = 2^k + j$*, where* $j < 2^k$ *and* $n \geq k$ *be arbitrary. Then,* $S[\mathcal{A}_i^{n+1}]$ *is isomorphic to* $\mathcal{A}_j^{k+1}$*, and* $S[\mathcal{B}_i^{n+1}]$ *is isomorphic to* $\mathcal{B}_j^{k+1}$.

Proof. By induction on n. In the base case, $n = k$, and $\mathcal{A}_i^{k+1}$ is defined by the clause for $i > 2^k$, from which it is obvious that $S[\mathcal{A}_i^{k+1}] \cong \mathcal{A}_j^{k+1}$. Similarly, $S[\mathcal{B}_i^{k+1}] \cong \mathcal{B}_j^{k+1}$.

If $n > k$, then $\mathcal{A}_i^{n+1}$ is defined by the clause for $i \leq 2^k$, meaning that it is a copy of $\mathcal{A}_i^n$ with an atom added to the root. It follows that

$$S[\mathcal{A}_i^{n+1}] \cong S[\mathcal{A}_i^n] \overset{\text{IH}}{\cong} S[\mathcal{A}_i^{k+1}],$$

as needed. The claim for $\mathcal{B}_i^{n+1}$ is analogous. □

In order for the Hydra to be allowed to set up the playing field with $\mathbf{A}^n \circ \mathbf{B}^n$, we need our formulas φ_n to be true on the left and false on the right, which is indeed the case. Intuitively, φ_n is made true by the rightmost branches of the models in $\mathbf{A}^n$, which is missing in $\mathbf{B}^n$. Note that these branches are pairwise different: a fact that is crucial for our subsequent arguments.

Lemma 7.9. *For all* $n \geq 0$, $\mathbf{A}^{n+1} \models \varphi_{n+1}$*, whereas* $\mathbf{B}^{n+1} \models \neg\varphi_{n+1}$.

Proof. By induction on n. If $n = 0$, then $\varphi_1 = \diamond p_1$, which is true on all pointed models of $\mathbf{A}^1$ and false on all models of $\mathbf{B}^1$, as can be seen by inspection on Figure 3.

If $n > 0$, recall that $\varphi_{n+1} = \diamond(p_{n+1} \wedge \diamond\varphi_n)$. Fix $i \leq 2^{n+1}$. Let $\boldsymbol{a} = (\mathcal{A}, a) = \boldsymbol{a}_i^{n+1} \in \mathbf{A}^n$ and $\boldsymbol{b} = (\mathcal{B}, b) = \boldsymbol{b}_i^{n+1} \in \mathbf{B}^{n+1}$. We consider two cases.

CASE $i \leq 2^n$. In this case, $\boldsymbol{a} = \boldsymbol{a}_i^{n+1} \vDash p_{n+1}$ by construction. By the induction hypothesis, $\boldsymbol{a}_i^n \vDash \varphi_n$. Since $\boldsymbol{a}$ agrees everywhere with $\boldsymbol{a}_i^n$ on all atoms different from p_{n+1}, it follows that $\boldsymbol{a} \vDash \varphi_n$ as well, and hence $\boldsymbol{a} \vDash p_{n+1} \wedge \oplus\varphi_n$, which readily implies that $\boldsymbol{a} \vDash \oplus(p_{n+1} \wedge \oplus\varphi_n)$.

As for $\boldsymbol{b} = \boldsymbol{b}_i^{n+1}$, by the induction hypothesis we have that $\boldsymbol{b}_i^n \nvDash \varphi_n$, which implies that $\boldsymbol{b} \nvDash \varphi_n$, since the two agree on all atoms appearing in φ_n. Now, consider arbitrary $v \in |\mathcal{B}|$ (so that $b \; R_{\mathcal{B}} \; v$). If we had that $(\mathcal{B}, v) \vDash \varphi_n$, by the transitivity of $R_{\mathcal{B}}$, we would have that $\boldsymbol{b} \vDash \varphi_n$, which is false. Hence, φ_n is false on every point of $|\mathcal{B}|$, from which it follows that $\oplus\varphi_n$ is false on every point of $|\mathcal{B}|$ as well. It follows that $\boldsymbol{b} \nvDash \oplus(p_{n+1} \wedge \oplus\varphi_n)$.

CASE $i > 2^n$. Write $i = 2^n + j$. Since we already have that $\boldsymbol{a}_j^n \vDash \varphi_{n+1}$ by the previous case, and $\boldsymbol{a}_j^n$ is locally bisimilar to $(\mathcal{A}, S_{\mathcal{A}}[a])$, we see that $\boldsymbol{a} \vDash \oplus\varphi_{n+1}$, which implies that $\boldsymbol{a} \vDash \varphi_{n+1}$.

Finally, for $\boldsymbol{b}$ we see by construction that the only point of $|\mathcal{B}|$ that satisfies p_{n+1} is $S_{\mathcal{B}}(b)$. However, by the previous case, $(\mathcal{B}, S_{\mathcal{B}}(b)) \nvDash \varphi_{n+1}$, from which it follows that $(\mathcal{B}, S_{\mathcal{B}}(b)) \nvDash \oplus\varphi_n$, and hence $\boldsymbol{b} \nvDash \varphi_{n+1}$. □

7.2 The lower bound on the number of moves in $(\varphi_n, \mathbf{GL})$-MEG

The pointed models in $\mathbf{A}^n \circ \mathbf{B}^n$ are constructed so that the critical branch of $\mathcal{A}_i^n$ is always very similar to the critical branch of $\mathcal{B}_i^n$, differing only at their top point (see, for example, Figure 6). Let us make this precise.

Definition 7.10. *Suppose that $\mathcal{M}$, $\mathcal{N}$ are two finite models with successors and with roots w and v, respectively. We say that $r \in \mathbb{N}$* distinguishes *$\mathcal{M}$ and $\mathcal{N}$ if $(\mathcal{M}, S_{\mathcal{M}}^r(w))$, $(\mathcal{N}, S_{\mathcal{N}}^r(v))$ differ on the truth of a propositional variable, but whenever $i < r$, then $(\mathcal{M}, S_{\mathcal{M}}^i(w))$, $(\mathcal{N}, S_{\mathcal{N}}^i(v))$ agree on the truth of all propositional variables. We call r the* distinguishing value *of $\mathcal{M}$ and $\mathcal{N}$.*

Note that the distinguishing value of two models $\mathcal{M}, \mathcal{N}$ need not be defined, but when it is, it is unique. Moreover, the distinguishing values of the models we have constructed usually do exist.

Lemma 7.11. *Fix $n \geq 1$ and $1 \leq i < j \leq 2^n$. Suppose that $\mathcal{A}_i^n$ and $\mathcal{A}_j^n$ have the same critical height m. Then, $\mathcal{A}_i^n$ and $\mathcal{A}_j^n$ are distinguished by some $r < m$, satisfying the following properties:*

(a) If $i \leq 2^{n-1}$ and $2^{n-1} < j$, then $\mathcal{A}_i^n$ and $\mathcal{A}_j^n$ have distinguishing value 0.

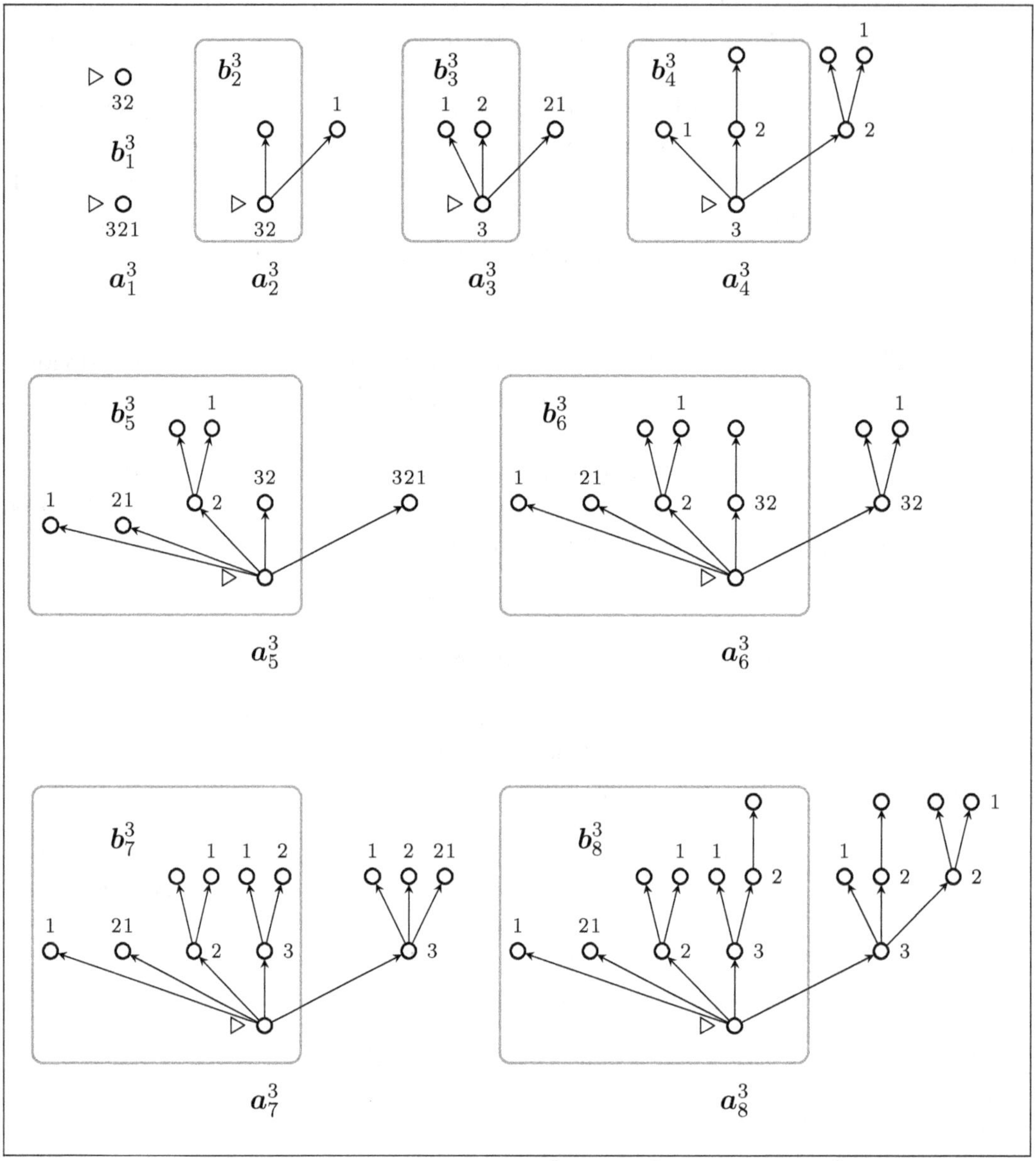

Figure 6: The pointed models in the sets $\mathbf{A}^3$ and $\mathbf{B}^3$. Note that $\boldsymbol{b}_j^i$ is a submodel of $\boldsymbol{a}_j^i$ obtained by deleting the rightmost branch, indicated in the figure within the gray boxes.

(b) *If $\mathcal{A}_i^n$ and $\mathcal{A}_j^n$ have distinguishing value r, then $\mathcal{A}_i^{n+1}$, $\mathcal{A}_j^{n+1}$ have distinguishing value r and $\mathcal{A}_{2^n+i}^{n+1}$, $\mathcal{A}_{2^n+j}^{n+1}$ have distinguishing value $r+1$.*

Proof. Assume that $\mathcal{A}_i^{n+1}$ and $\mathcal{A}_j^{n+1}$ are so that their critical heights have the same value, m. To prove (a), it suffices to observe that in this case, the root of $\mathcal{A}_i^{n+1}$ satisfies p_{n+1}, but not the root of $\mathcal{A}_j^{n+1}$. For (b), if $\mathcal{A}_i^n$ and $\mathcal{A}_j^n$ are distinguished by r, it is easy to see that $\mathcal{A}_i^{n+1}$ and $\mathcal{A}_j^{n+1}$ are still distinguished by r, since the models agree on all variables p_k with $k \leq n$, and p_{n+1} is true exactly on both roots of the new models. Meanwhile, $S[\mathcal{A}_{2^n+i}^{n+1}] \cong \mathcal{A}_i^{n+1}$ and $S[\mathcal{A}_{2^n+j}^{n+1}] \cong \mathcal{A}_j^{n+1}$, hence $\mathcal{A}_i^{n+1}$ and $\mathcal{A}_j^{n+1}$ are distinguished by $r+1$ since we have added a new root to each model, both satisfying no atoms.

From this, the existence of a distinguishing value $r < m$ follows by a straightforward induction on n. The base case, $n = 1$, follows vacuously since no two models in $\mathbf{A}^1$ have the same critical height. For the inductive case, we assume the claim for n and prove it for $n+1$. If $i, j \leq 2^n$, then by the induction hypothesis we have that $\mathcal{A}_i^n$ and $\mathcal{A}_j^n$ are distinguished by some $r < m$, and by (b), $\mathcal{A}_i^{n+1}$ and $\mathcal{A}_j^{n+1}$ are still distinguished by r. If $i \leq 2^n$ and $2^n < j$, by (a), $\mathcal{A}_i^{n+1}$ and $\mathcal{A}_j^{n+1}$ are distinguished by $r = 0$. Finally, if $2^n < i, j$, by the induction hypothesis, $\mathcal{A}_{i-2^n}^n$ and $\mathcal{A}_{j-2^n}^n$ are distinguished by some $r < m-1$, so that by (b), $\mathcal{A}_i^{n+1}$ and $\mathcal{A}_j^{n+1}$ are distinguished by $r+1 < m$.

□

Lemma 7.12. *Fix $n \geq 1$ and $i \in [1, 2^n]$. Then, $\mathcal{A}_i^n$ and $\mathcal{B}_i^n$ have the same critical height m, and are distinguished by m.*

Proof. Proceed by induction on n. The base case follows from inspecting $\mathbf{A}^1, \mathbf{B}^1$, depicted in Figure 3. For the inductive step, we assume the claim for n and prove it for $n+1$. If $i \leq 2^n$, since $\boldsymbol{A}_i^{n+1}$ and $\boldsymbol{B}_i^{n+1}$ are based on the same frames as $\boldsymbol{A}_i^n$ and $\boldsymbol{B}_i^n$, respectively, it follows from the induction hypothesis that they share the same critical height, m. Then, for $k < m$, $S^k[\boldsymbol{a}_i^n]$ and $S^k[\boldsymbol{b}_i^n]$ agree on all atoms by the induction hypothesis, and hence $S^k[\boldsymbol{a}_i^{n+1}]$ and $S^k[\boldsymbol{b}_i^{n+1}]$ agree on all atoms too, including p_{n+1}, which is true precisely on the roots of both models. Moreover, $S^m[\boldsymbol{a}_i^n]$ and $S^m[\boldsymbol{b}_i^n]$ disagree on some atom (in fact, on p_1), hence so too do $S^m[\boldsymbol{a}_i^{n+1}]$ and $S^m[\boldsymbol{b}_i^{n+1}]$.

If $2^n < i$, then $S[\boldsymbol{a}_i^{n+1}] \cong \boldsymbol{a}_{i-2^n}^{n+1}$ and $S[\boldsymbol{b}_i^{n+1}] \cong \boldsymbol{b}_{i-2^n}^n$, which by the previous case share the same critical height, m. It follows that $\boldsymbol{a}_i^{n+1}$ and $\boldsymbol{b}_i^{n+1}$ also share the same critical height, $m+1$. Moreover, $\boldsymbol{a}_i^{n+1}$ and $\boldsymbol{b}_i^{n+1}$ agree on all atoms (they are all false), and since $S^{r+1}[\boldsymbol{a}_i^{n+1}] \cong S^r[\boldsymbol{a}_i^n]$ and $S^{r+1}[\boldsymbol{b}_i^{n+1}] \cong S^r[\boldsymbol{b}_i^n]$ for $r \leq m$, it follows once again by the previous case that they share the same atoms for $r < m$ and disagree on some atom for $r = m$. □

By *twins of height* k we mean a pair of the form $(S^k[\boldsymbol{a}_i^n], S^k[\boldsymbol{b}_i^n])$, where $i \leq 2^n$ and both expressions are defined. If $\mathbf{L}$ is a set of pointed models from $\mathbf{A}^n$ and $\mathbf{R}$ from $\mathbf{B}^n$, we say that there are twins of height k in $\mathbf{L} \circ \mathbf{R}$ if there are twins $(S^k[\boldsymbol{a}_i^n], S^k[\boldsymbol{b}_i^n])$ such that $S^k[\boldsymbol{a}_i^n] \in \mathbf{L}$ and $S^k[\boldsymbol{b}_i^n] \in \mathbf{R}$.

For example, the pointed models $(\boldsymbol{a}_{2^n+1}^{n+1}, \boldsymbol{b}_{2^n+1}^{n+1})$ shown in Figure 4 are twins of height zero, and $(S[\boldsymbol{a}_{2^n+1}^{n+1}], S[\boldsymbol{b}_{2^n+1}^{n+1}])$ are twins of height one. Note that the two pairs share the same models, and vary only on the evaluation point: $\boldsymbol{a}_{2^n+1}^{n+1}$ and $\boldsymbol{b}_{2^n+1}^{n+1}$ are evaluated at their respective roots, $S[\boldsymbol{a}_{2^n+1}^{n+1}]$ and $S[\boldsymbol{b}_{2^n+1}^{n+1}]$ at the rightmost daughters of these roots.

The following lemma tells us that, while models $\mathcal{A}_i^n$ and $\mathcal{B}_j^n$ can never be bisimilar (since one satisfies φ_n but the other does not), they can come quite close to being so. Recall that if $\boldsymbol{a}, \boldsymbol{b}$ are pointed models, we write $\boldsymbol{a} \leftrightarroows \boldsymbol{b}$ to indicate that they are locally bisimilar.

Lemma 7.13. *Fix $n \geq 1$ and $r \geq 0$.*

(i) If $(\boldsymbol{a}, \boldsymbol{b})$ are twins of height r in $\mathbf{A}^n \circ \mathbf{B}^n$ and $\boldsymbol{a}' \in \Box \boldsymbol{a}$ is such that $\boldsymbol{a}' \not\leftrightarroows S[\boldsymbol{a}]$, then there is $\boldsymbol{b}' \in \Box \boldsymbol{b}$ such that $\boldsymbol{a}' \leftrightarroows \boldsymbol{b}'$

(ii) If $(\boldsymbol{a}, \boldsymbol{b})$ are twins of height r in $\mathbf{A}^n \circ \mathbf{B}^n$ and $\boldsymbol{b}' \in \Box \boldsymbol{b}$, then there is $\boldsymbol{a}' \in \Box \boldsymbol{a}$ such that $\boldsymbol{a}' \leftrightarroows \boldsymbol{b}'$.

(iii) If $1 \leq i < j \leq 2^n$, $\mathcal{A}_i^n$ and $\mathcal{B}_j^n$ have the same critical height $m > r$, and they are distinguished by r, then there is $\boldsymbol{b}' \in \Box S^r[\boldsymbol{b}_j^n]$ such that $S^{r+1}[\boldsymbol{a}_i^n] \leftrightarroows \boldsymbol{b}'$.

Proof. The three claims are proven by induction on n.

(i). The base case (for $n = 1$) follows by observing Figure 3. Indeed, the only twins are $(\boldsymbol{a}_1^1, \boldsymbol{b}_1^1)$, $(\boldsymbol{a}_2^1, \boldsymbol{b}_2^1)$ and $(S[\boldsymbol{a}_2^1], S[\boldsymbol{b}_2^1])$. Of these, only for $(\boldsymbol{a}_2^1, \boldsymbol{b}_2^1)$ can we choose $\boldsymbol{a}'$ as in the antecedent, so in the other cases the claim is vacuously true. But in this case, we have that $\boldsymbol{a}'$ must be $\mathcal{A}_2^1$ evaluated at the top-left point, which is clearly locally bisimilar to $\mathcal{B}_2^1$ evaluated at its top point.

Otherwise, assume the claim for n, and let us establish it for $n + 1$. Write $\boldsymbol{a} = (\mathcal{A}, a)$ and $\boldsymbol{b} = (\mathcal{B}, b)$, and assume that $i \in [1, 2^{n+1}]$ is so that

$$(\boldsymbol{a}, \boldsymbol{b}) = (S^r[\boldsymbol{a}_i^{n+1}], S^r[\boldsymbol{b}_i^{n+1}]).$$

Let $\boldsymbol{a}' = (\mathcal{A}, a') \in \Box \boldsymbol{a}$ be such that $\boldsymbol{a}' \not\leftrightarroows S[\boldsymbol{a}]$, and consider two cases, according to i.

Case $i \leq 2^n$**.** Observe that a' cannot be the root of $\mathcal{A}$, since $R_{\mathcal{A}}$ is irreflexive. It follows that $(\mathcal{A}, a') \leftrightarrows (\mathcal{A}_i^n, a')$ since, by definition, $\mathcal{A}_i^n$ and $\mathcal{A} = \mathcal{A}_i^{n+1}$ disagree only at the root. By the induction hypothesis, there is $b' \in |\mathcal{B}_i^n|$ such that $b \; R_{\mathcal{B}} \; b'$ and $(\mathcal{A}_i^n, a') \leftrightarrows (\mathcal{B}_i^n, b')$, but once again b' cannot be the root so we must have that $(\mathcal{B}_i^n, b') \leftrightarrows (\mathcal{B}, b') \in \Box\boldsymbol{b}$, as claimed.

Case $i > 2^n$**.** We consider two sub-cases. First, assume that a is *not* the root of $\mathcal{A}$, so that $a \in |S[\mathcal{A}]|$. In this case, by Lemma 7.8, $(\mathcal{A}, a) \leftrightarrows (\mathcal{A}_{i-2^n}^n, a)$ and $(\mathcal{B}, b) \leftrightarrows (\mathcal{B}_{i-2^n}^n, b)$. We can then apply the case for $i \leq 2^n$ to find $b' \in |S[\mathcal{B}]|$ such that $b \; R_{\mathcal{B}} \; b'$ and $(S[\mathcal{B}], b') \leftrightarrows (S[\mathcal{A}], a')$. This gives us that $(\mathcal{B}, b') \leftrightarrows (\mathcal{A}, a')$ as well.

Otherwise, assume that a is the root of $\mathcal{A}$, so that b is also the root of $\mathcal{B}$. If $a' \in \left|\coprod_{j=2}^{2^n} S[\mathcal{A}_j^{n+1}] \sqcup \mathcal{B}_j^{n+1}\right|$, we can take $b' = a'$, and clearly $(\mathcal{A}, a')$ is locally bisimilar to $(\mathcal{B}, b')$. Otherwise, $a' \in |S[\mathcal{A}]|$, and by the assumption $a' \neq S_{\mathcal{A}}(a)$.

If $a' = S_{\mathcal{A}}(S_{\mathcal{A}}(a))$, we let b' be the copy of $S_{\mathcal{A}}(S_{\mathcal{A}}(a))$ in $\coprod_{j=2}^{2^n} S[\mathcal{A}_j^{n+1}]$. If not, since $S_{\mathcal{A}}(a)$ is the root of $S[\mathcal{A}]$, we have that $S_{\mathcal{A}}(a) \; R_{\mathcal{A}} \; a'$. Since $S[\boldsymbol{a}] \leftrightarrows \boldsymbol{a}_{i-2^n}^{n+1}$, we can apply the case for $i \leq 2^n$ to find $b' \in |S[\mathcal{B}]|$ such that $S_{\mathcal{B}}(b) \; R_{\mathcal{B}} \; b'$ and $(\mathcal{B}, b') \leftrightarrows (\mathcal{A}, a')$. By transitivity we also have that $b \; R_{\mathcal{B}} \; b'$, as needed.

(ii). The base case can readily be verified for $\mathbf{A}^1 \circ \mathbf{B}^1$ on Figure 3. The inductive step follows the same structure as that for claim (i) by swapping the roles of $\boldsymbol{a}$ and $\boldsymbol{b}$, except that in the case where $i > 2^n$ and b is the root of $\mathcal{B}$, the proof is somewhat simplified as we always have $b' \in \left|\coprod_{j=2}^{2^n} S[\mathcal{A}_j^{n+1}] \sqcup \mathcal{B}_j^{n+1}\right|$, and thus we can always take $a' = b'$.

(iii). It is obvious that the proposition is trivially true for $n = 1$ because the respective critical heights of $\mathcal{A}_1^1$ and $\mathcal{B}_2^1$ are different. The inductive step also follows a similar structure as before, but in order to apply the induction hypothesis we must also pay some attention to the distinguishing values.

Case $i, j \leq 2^n$**.** By Lemma 7.11(b), if $\mathcal{A}_i^{n+1}$ and $\mathcal{B}_j^{n+1}$ are distinguished by r, then so are $\mathcal{A}_i^n$ and $\mathcal{B}_j^n$, which by the induction hypothesis tells us that there is $(\mathcal{B}_j^n, b') \in \Box S^r[\boldsymbol{b}_j^n]$ that is locally bisimilar to $S^{r+1}[\boldsymbol{a}_i^n]$. Reasoning as in the case for $i \leq 2^n$ in claim (i), this yields $S^{r+1}[\boldsymbol{a}_i^{n+1}] \leftrightarrows (\mathcal{B}_j^{n+1}, b') \in \Box S^r[\boldsymbol{b}_j^{n+1}]$.

Case $i, j > 2^n$**.** We once again use Lemma 7.11(b) (twice) to see that if $\mathcal{A}_i^{n+1}$ and $\mathcal{B}_j^{n+1}$ are distinguished by r, then $\mathcal{A}_{i-2^n}^{n+1}$ and $\mathcal{B}_{j-2^n}^{n+1}$ are distinguished by $r-1$. By the case for $i, j \leq 2^n$, this tells us that there is $(\mathcal{B}_{j-2^n}^{n+1}, b') \in \Box S^{r-1}[\boldsymbol{b}_{j-2^n}^{n+1}]$ that is locally bisimilar to $S^r[\boldsymbol{a}_{i-2^n}^{n+1}]$. Setting $\boldsymbol{b}' = (\mathcal{B}_j^{n+1}, b')$, we reason as in the proof of the case

$i > 2^n$ in claim (i) to obtain $S^{r+1}[\boldsymbol{a}_i^{n+1}] \leftrightarroweq \boldsymbol{b}'$ and $\boldsymbol{b}' \in \Box S^r[\boldsymbol{b}_j^{n+1}]$.

Case $i \leq 2^n < j$. By Lemma 7.11(a), $\mathcal{A}_i^{n+1}$ and $\mathcal{B}_j^{n+1}$ are distinguished by $r = 0$. But, $\coprod_{k=2}^{2^n} S[\mathcal{A}_k^{n+1}]$ already contains a copy of $S[\boldsymbol{a}_i^{n+1}]$, and we use this copy as $\boldsymbol{b}'$. □

Example 7.14. *Lemma 7.13(iii) is best understood by looking at the models in Figures 5 and 6. A simple inspection of Figure 5 is enough to see that $\boldsymbol{a}_2^2$ and $\boldsymbol{b}_3^2$ differ on the truth of p_2 and that $S[\boldsymbol{a}_2^2]$ is locally bisimilar to $\mathcal{B}_3^2$ at its top-left point. Meanwhile, in Figure 6, $\boldsymbol{a}_2^3$ and $\boldsymbol{b}_3^3$ differ on the value of p_2, i.e., the smallest number for which the critical branches of $\mathcal{A}_2^3$ and $\mathcal{B}_3^3$ differ on the truth of a propositional variable is zero. Obviously, $S[\boldsymbol{a}_2^3]$ satisfies only p_1 and the same applies to the* left *successor point of $\boldsymbol{b}_3^3$. In a similar way, we see that the smallest number for which the critical branches of $\mathcal{A}_6^3$ and $\mathcal{B}_7^3$ differ on the truth of a propositional variable is one because the rightmost daughters s and t of the roots of $\mathcal{A}_6^3$ and $\mathcal{B}_7^3$, respectively, differ on the truth of p_2. Again, we have that the rightmost daughter of s satisfies only p_1 and the same applies to the left successor of the only node in $\mathcal{B}_7^3$ that satisfies p_3.*

Lemma 7.13 shows us that the moves that Hercules can make in order to win are, in fact, rather restricted. Below, for fixed $n \geq 1$, say that the Hydra *plays well* if she labels the root by $\mathbf{A}^n \circ \mathbf{B}^n$ and plays greedily.

Lemma 7.15. *Assume that the Hydra plays well. For any node η (not necessarily a leaf) in a closed game tree T for the $(\varphi_n, \mathbf{GL})$-MEG:*

(a) if there are twins $(\boldsymbol{a}, \boldsymbol{b})$ in $\mathfrak{L}(\eta) \circ \mathfrak{R}(\eta)$, then Hercules did not play a □-move in η;

(b) if there are twins $(\boldsymbol{a}, \boldsymbol{b})$ in $\mathfrak{L}(\eta) \circ \mathfrak{R}(\eta)$ and Hercules played a ◇-move, then he chose $S[\boldsymbol{a}] \in \Box\boldsymbol{a}$, and

(c) if there there are two pairs of twins $(\boldsymbol{a}, \boldsymbol{b})$ and $(\boldsymbol{a}', \boldsymbol{b}')$ both of height r in $\mathfrak{L}(\eta) \circ \mathfrak{R}(\eta)$ and r distinguishes $\boldsymbol{a}$ and $\boldsymbol{a}'$, then Hercules did not play a ◇-move at η.

Proof. Assume that Hercules played either a ◇-move or a □-move, and let η' be the new head that was created.

(a). If $\mathfrak{L}(\eta) \circ \mathfrak{R}(\eta)$ contains twins $(\boldsymbol{a}, \boldsymbol{b})$ and Hercules plays a ◇-move in η, he must choose $\boldsymbol{a}' \in \Box\boldsymbol{a}$ to place in $\mathfrak{L}(\eta')$. If $\boldsymbol{a}' \neq S[\boldsymbol{a}]$, then by Lemma 7.13(i), there is $\boldsymbol{b}' \in \Box\boldsymbol{b}$ such that $\boldsymbol{a}' \leftrightarroweq \boldsymbol{b}'$. Since the Hydra plays greedily, we have that $\boldsymbol{b}' \in \mathfrak{R}(\eta')$, which by Lemma 6.6 implies that Hercules cannot win.

(b). This is simliar to the previous item. If Hercules plays a $\Box$-move in η, then he must choose $\boldsymbol{b}' \in \Box\boldsymbol{b}$ to place in $\mathfrak{L}(\eta')$. But then, by Lemma 7.13(ii), the Hydra will place a bisimilar $\boldsymbol{a}' \in \mathfrak{R}(\eta')$, and Hercules cannot win.

(c). Assume that $(S^r[\boldsymbol{a}_i^n], S^r[\boldsymbol{b}_i^n])$ and $(S^r[\boldsymbol{a}_j^n], S^r[\boldsymbol{b}_j^n])$ are two pairs of twins in $\mathfrak{L}(\eta) \circ \mathfrak{R}(\eta)$ with $i < j$, and such that r distinguishes $\mathcal{A}_i^n$ and $\mathcal{A}_j^n$. If Hercules plays a $\Box$-move, by claim (a), he must place $S^{r+1}[\boldsymbol{a}_i^n] \in \mathfrak{L}(\eta')$. By Lemma 7.13(iii), there will be $\boldsymbol{v} \in \Box S^r[\boldsymbol{b}_j^n]$ such that $S^{r+1}[\boldsymbol{a}_i^n] \leftrightarroweq \boldsymbol{v}$. As before, this causes there to be bisimilar pointed models in $\mathfrak{L}(\eta')$ and $\mathfrak{R}(\eta')$, which implies that Hercules cannot win. □

Since the respective rightmost branches in the pointed models $\boldsymbol{a}_j^n$ and $\boldsymbol{b}_j^n$ differ on a literal only in their leaves, we see that for every pair of twins $(\boldsymbol{a}, \boldsymbol{b})$, the Hydra's strategy forces Hercules to make m many $\Diamond$-moves, where m is the critical height of $\boldsymbol{a}$ and $\boldsymbol{b}$. Let us make this precise.

Definition 7.16. *Fix $n \geq 1$, $i \in [1, 2^n]$ and a closed game tree $(T, \preccurlyeq)$. Then, define $\Lambda(i)$ to be the set of leaves η of T such that for every $\eta' \preccurlyeq \eta$, there is some $r \geq 0$ such that $(S^r[\boldsymbol{a}_i^n], S^r[\boldsymbol{b}_i^n])$ appear in $\mathfrak{L}(\eta') \circ \mathfrak{R}(\eta')$.*

The sets $\Lambda(i)$ are non-empty and disjoint when Hydra plays well, from which our exponential lower bound will follow immediately. To prove this, we will need the following lemma.

Lemma 7.17. *Fix $n \geq 1$. Let T be a closed game-tree for the $(\varphi_n, \mathbf{GL})$-MEG where the Hydra plays well. Let $i \in [1, 2^n]$, and $\eta \in \Lambda(i)$.*

(a) *For all $\zeta \preccurlyeq \eta$ and all $r \geq 0$, if Hercules has played r $\Diamond$-moves before ζ then $(S^r[\boldsymbol{a}_i^n], S^r[\boldsymbol{b}_i^n])$ appear in $\mathfrak{L}(\zeta) \circ \mathfrak{R}(\zeta)$.*

(b) *If Hercules played m $\Diamond$-moves before η, then m is the critical height of $\mathcal{A}_i^n$ and $\mathcal{B}_i^n$.*

Proof. Fix T as in the statement of the lemma, and let η_0 be its root.

(a). We proceed by induction on ζ along $\prec$. For the induction to go through, we need to prove a slightly stronger claim: if Hercules has played r $\Diamond$-moves before ζ, then $(S^r[\boldsymbol{a}_i^n], S^r[\boldsymbol{b}_i^n])$ appear in $\mathfrak{L}(\zeta) \circ \mathfrak{R}(\zeta)$, and

(c) for all $t \neq r$, $S^t[\boldsymbol{a}_i^n] \notin \mathfrak{L}(\zeta)$.

For the base case this is clear, as only $S^0[\boldsymbol{a}_i^n], S^0[\boldsymbol{b}_i^n]$ appear in $\mathbf{A}^n \circ \mathbf{B}^n = \mathfrak{L}(\eta_0) \circ \mathfrak{R}(\eta_0)$, and Hercules has played zero $\Diamond$-moves before η_0.

For the inductive step, assume the claim for ζ, and suppose that $\zeta' \preccurlyeq \eta$ is a daughter of ζ; we will prove claims (a) and (c) for ζ'. Let r be the number of $\Diamond$-moves that Hercules has played before ζ. Since $\eta \in \Lambda(i)$, we have that $(S^k[\boldsymbol{a}_i^n], S^k[\boldsymbol{b}_i^n])$ occur in $\mathfrak{L}(\zeta') \circ \mathfrak{R}(\zeta')$ for some k.

Hercules obviously did not play a literal move on ζ, or it would be a leaf. If Hercules played a $\vee$- or $\wedge$-move, since these moves do not introduce new pointed models, it follows that $(S^k[\boldsymbol{a}_i^n], S^k[\boldsymbol{b}_i^n])$ also occur in $\mathfrak{L}(\zeta) \circ \mathfrak{R}(\zeta)$, and by uniqueness that $k = r$, from which claim (a) follows for ζ'. As for claim (c), if $S^t[\boldsymbol{a}_i^n] \in \mathfrak{L}(\zeta')$, then once again we have that $S^t[\boldsymbol{a}_i^n] \in \mathfrak{L}(\zeta)$ and thus $t = r$.

If Hercules played a $\Diamond$-move, then $(S^r[\boldsymbol{a}_i^n], S^r[\boldsymbol{b}_i^n])$ occur in $\mathfrak{L}(\zeta) \circ \mathfrak{R}(\zeta)$ by the induction hypothesis. By Lemma 7.15(i), Hercules chose $S^{r+1}[\boldsymbol{a}_i^n] \in \Box S^r[\boldsymbol{a}_i^n]$. By Lemma 7.12, $\mathcal{A}_i^n$ has the same critical height as $\mathcal{B}_i^n$, and thus $S^{r+1}[\boldsymbol{b}_i^n]$ is defined. Since the Hydra plays greedily, she chose $S^{r+1}[\boldsymbol{b}_i^n] \in \Box S^r[\boldsymbol{b}_i^n]$. But, there are now $r+1$ $\Diamond$-moves before ζ', so claim (a) follows. Moreover, if $S^t[\boldsymbol{a}_i^n] \in \mathfrak{L}(\zeta')$, then there must be $\boldsymbol{a}' \in \mathfrak{L}(\zeta)$ such that $S^t[\boldsymbol{a}_i^n] \in \Box\boldsymbol{a}'$. But, since $\mathcal{A}_i^n$ is a tree, this can only occur when $\boldsymbol{a}' = S^{t'}[\boldsymbol{a}_i^n]$ for some $t' < t$, and it follows that $t' = r$ by the induction hypothesis, so that once again by Lemma 7.15(i), $t = r + 1$. Claim (c) follows.

Finally, we note that Hercules cannot play a $\Box$-move on ζ by Lemma 7.15(b).

(b). Let r be the number of $\Diamond$-moves that Hercules played before η. By Lemma 7.12, if m is the critical height of $\mathcal{A}_i^n$, then it is also the critical height of $\mathcal{B}_i^n$ and m distinguishes $\mathcal{A}_i^n$ and $\mathcal{B}_i^n$. Since T is closed, η must be a stub, which means that Hercules must have played a literal move on η. But this is only possible if $S^r[\boldsymbol{a}_i^n]$ and $S^r[\boldsymbol{b}_i^n]$ disagree on a literal, which is only possible if $r = m$. □

Lemma 7.18. *Fix $n \geq 1$. Let $(T, \preccurlyeq)$ be a closed game-tree for the $(\varphi_n, \mathbf{GL})$-MEG where the Hydra plays well.*

(a) For all $i \in [1, 2^n]$, $\Lambda(i)$ is non-empty.

(b) If $1 \leq i < j \leq 2^n$, then $\Lambda(i) \cap \Lambda(j) = \varnothing$.

Proof. Let $\boldsymbol{a} = \boldsymbol{a}_i^n$ and $\boldsymbol{b} = \boldsymbol{b}_i^n$.

(a) We show by induction on the number of rounds in the game that there is always a leaf η such that

(∗) for all $\zeta \preccurlyeq \eta$ there is $r \geq 0$ such that $(S^r[\boldsymbol{a}], S^r[\boldsymbol{b}])$ appear in $\mathfrak{L}(\zeta) \circ \mathfrak{R}(\zeta)$.

For the base case, we take η to be the root, in which case it is clear that $(S^0[\boldsymbol{a}], S^0[\boldsymbol{b}])$ appears in $\mathfrak{L}(\eta) \circ \mathfrak{R}(\eta) = \mathbf{A}^n \circ \mathbf{B}^n$. For the inductive step, assume that η is a leaf such that $(*)$ holds. We may assume that Hercules plays on η, for otherwise η remains on the game-tree as a leaf.

If Hercules plays a literal move, then η simply becomes a stub, but remains on the game-tree. If Hercules plays a $\vee$- or $\wedge$-move, then two heads η_1 and η_2 are added, and as in the proof of Lemma 6.6, either $(S^r[\boldsymbol{a}], S^r[\boldsymbol{b}])$ occurs in $\mathfrak{L}(\eta_1) \circ \mathfrak{R}(\eta_1)$ and we take η_1 as the new head, or it occurs in $\mathfrak{L}(\eta_2) \circ \mathfrak{R}(\eta_2)$ and we take η_2 instead.

If Hercules plays a $\diamond$-move, then a new node η' is added, and by Lemma 7.15(i), Hercules places $S^{r+1}[\boldsymbol{a}]$ in $\mathfrak{L}(\eta')$. Since the Hydra plays greedily and $S^{r+1}[\boldsymbol{b}]$ exists by Lemma 7.12, we have that $S^{r+1}[\boldsymbol{b}] \in \mathfrak{R}(\eta')$. Therefore, the twins $(S^{r+1}[\boldsymbol{a}], S^{r+1}[\boldsymbol{b}])$ appear in $\mathfrak{L}(\eta') \circ \mathfrak{R}(\eta')$. Finally, Hercules cannot play a $\Box$-move by Lemma 7.15(b).

(b) Now, let $1 \leq i < j \leq 2^n$. Towards a contradiction, assume that $\eta \in \Lambda(i) \cap \Lambda(j)$. Let m be the number of $\diamond$-moves that Hercules played before η. By Lemma 7.17(b), $\mathcal{A}_i^n$ and $\mathcal{A}_j^n$ both have critical height m. By Lemma 7.11, there is $r < m$ that distinguishes $\mathcal{A}_i^n$ and $\mathcal{A}_j^n$. Let $\zeta' \preccurlyeq \eta$ be the first node such that Hercules has played $r+1$ $\diamond$-moves before ζ', and ζ be its predecessor. Then, by Lemma 7.17(a), $(S^r[\boldsymbol{a}_i^n], S^r[\boldsymbol{b}_i^n])$ and $(S^r[\boldsymbol{a}_j^n], S^r[\boldsymbol{b}_j^n])$ both appear on $\mathfrak{L}(\zeta) \circ \mathfrak{R}(\zeta)$, which by Lemma 7.15(c) implies that Hercules cannot play a $\diamond$-move at ζ. This means that he cannot have played $r+1$ $\diamond$-moves before ζ', a contradiction. □

We are finally ready to prove our lower bound on the number of moves in the $(\varphi_n, \mathbf{GL})$-MEG. In fact, we have proven a slightly stronger claim.

Proposition 7.19. *For every $n \geq 1$, Hercules has no winning strategy of less than 2^n moves in the $(\varphi_n, \mathbf{A}^n \cup \mathbf{B}^n)$-MEG.*

Proof. Assume that Hydra plays well, and let T be a closed game tree. Then, by Lemma 7.18, the sets of leaves $\{\Lambda(i) : i \in [1, 2^n]\}$ are non-empty and disjoint. It follows that there are at least 2^n leaves, and since closing each leaf requires one literal move, Hercules must have played at least 2^n moves. □

Since $\mathbf{A}^n \cup \mathbf{B}^n \subseteq \mathbf{GL}$, Theorem 7.3(a) readily follows. In view of Theorem 6.4, we also obtain the following stronger form of Proposition 7.4.1:

Proposition 7.20. *For all $n \geq 1$, whenever $\psi \in \mathcal{L}_\diamond$ is such that $\varphi_n \equiv \psi$ on $\mathbf{A}^n \cup \mathbf{B}^n$, it follows that $|\psi| \geq 2^n$.*

7.3 $\mathcal{L}_{\oplus}$ is exponentially more succinct than $\mathcal{L}_{\diamond}$ on TC

We proceed to show that Hercules has no winning strategy of less than 2^n moves in $(\varphi_n, \mathbf{TC})$-MEG. We begin by defining two sets of pointed models $\hat{\mathbf{A}}^n$ and $\hat{\mathbf{B}}^n$ that are a slight modification of the models in $\mathbf{A}^n$ and $\mathbf{B}^n$, respectively.

Definition 7.21. *Let $\mathcal{K} = (|\mathcal{K}|, R_{\mathcal{K}}, V_{\mathcal{K}})$ be any Kripke model. We define a new model $\hat{\mathcal{K}}$ such that*

(i) $|\hat{\mathcal{K}}| = |\mathcal{K}| \cup \{\infty\}$, where $\infty \notin |\mathcal{K}|$,

(ii) $R_{\hat{\mathcal{K}}} = R_{\mathcal{K}} \cup (|\hat{\mathcal{K}}| \times \{\infty\})$, and

(iii) $V_{\hat{\mathcal{K}}}(w) = V_{\mathcal{K}}(w)$ if $w \in |\mathcal{K}|$, $V_{\hat{\mathcal{K}}}(\infty) = \varnothing$.

If $\mathcal{K}$ is equipped with a successor partial function $S_{\mathcal{K}}$, we also define $S_{\hat{\mathcal{K}}} = S_{\mathcal{K}}$. For a class of models $\mathbf{X}$, we denote $\{\hat{\mathcal{K}} : \mathcal{K} \in \mathbf{X}\}$ by $\hat{\mathbf{X}}$.

In other words, we add a 'point at infinity' that is seen by both worlds. Note that the successor function remains unchanged, i.e. ∞ is never a successor. In particular, ∞ can never belong to a critical branch.

This operation allows us to easily turn a model into a totally connected model:

Lemma 7.22. *Let $\mathcal{M}, \mathcal{N}$ be* **K4** *models. Then:*

(a) *$\hat{\mathcal{M}}$ is a* **TC** *model;*

(b) *if $w \in |\mathcal{M}|$ and $v \in |\mathcal{N}|$ are such that $(\mathcal{M}, w) \leftrightarrows (\mathcal{N}, v)$, then $(\hat{\mathcal{M}}, w) \leftrightarrows (\hat{\mathcal{N}}, v)$, and*

(c) *$(\hat{\mathcal{M}}, \infty) \leftrightarrows (\hat{\mathcal{N}}, \infty)$.*

Proof. If $w, w' \in |\hat{\mathcal{M}}|$, then w, ∞, w' is a path connecting w to w' (as $x \, R_{\hat{\mathcal{M}}} \, \infty$ holds for all $x \in |\hat{\mathcal{M}}|$). It follows that $\hat{\mathcal{M}}$ is connected, and indeed the same path witnesses that $\hat{\mathcal{M}}$ is locally connected if we take $w, w' \in R_{\mathcal{M}}(u)$. The second claim follows from observing that if $\chi \subseteq |\mathcal{M}| \times |\mathcal{N}|$ is a bisimulation then so is $\hat{\chi} = \chi \cup \{(\infty, \infty)\} \subseteq |\hat{\mathcal{M}}| \times |\hat{\mathcal{N}}|$, and the third by choosing an arbitrary such $\hat{\chi}$ (setting $\chi = \varnothing$, so that $\hat{\chi} = \{(\infty, \infty)\}$, will do). □

The following analogue of Lemma 7.9 then holds; we omit the proof, which is identical.

Lemma 7.23. *For any $n \geq 0$, $\hat{\mathbf{A}}^{n+1} \models \varphi_{n+1}$ and $\hat{\mathbf{B}}^{n+1} \models \neg\varphi_{n+1}$.*

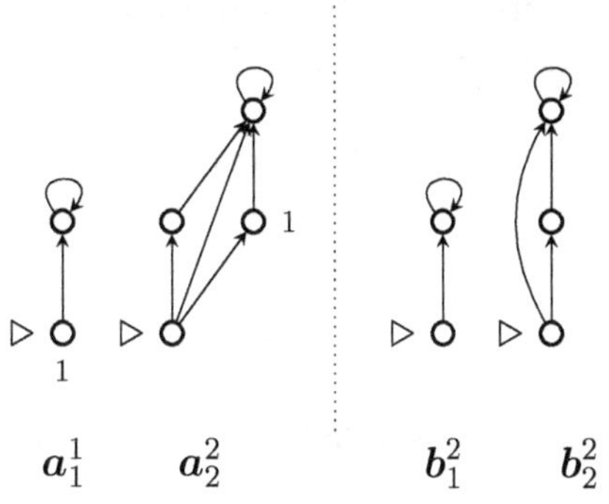

Figure 7: The pointed models in $\hat{\mathbf{A}}^1$ and $\hat{\mathbf{B}}^1$ are shown on the left and on the right of the dotted line, respectively.

Intuitively, the proof that Hercules has no winning strategy of less than 2^n moves in the game that starts with $\hat{\mathbf{A}}^n \circ \hat{\mathbf{B}}^n$ revolves around the observation that the models in $\hat{\mathbf{A}}^n$ and $\hat{\mathbf{B}}^n$ are constructed in such a way that Hercules, when playing a $\Diamond$- or $\Box$-move at some position labeled by $\mathbf{L} \circ \mathbf{R}$, cannot pick ∞ in any model of $\mathbf{L}$ or $\mathbf{R}$, as this will lead to bisimilar pointed models on each side. Thus, given the construction of the models in $\hat{\mathbf{A}}^n$ and $\hat{\mathbf{B}}^n$, Hercules and the Hydra are essentially playing a $(\varphi_n, \mathbf{GL})$-MEG on $\mathbf{A}^n \circ \mathbf{B}^n$ and the lower bound on the number of moves in any winning $(\varphi_n, \mathbf{GL})$-MEG for Hercules on $\mathbf{A}^n \circ \mathbf{B}^n$ established in the previous sub-section applies to the present case too.

Let us formalise the above intuitive considerations. As before, we will say that the Hydra plays well in the $(\varphi_n, \mathbf{TC})$-MEG if she labels the root by $\hat{\mathbf{A}}^n \circ \hat{\mathbf{B}}^n$ and plays greedily.

Lemma 7.24. *Fix $n \geq 1$, and assume that the Hydra plays well in the $(\varphi_n, \mathbf{TC})$-MEG. Then, for any position η in a closed game tree T such that $\mathfrak{L}(\eta)$ and $\mathfrak{R}(\eta)$ are both non-empty,*

(i) if Hercules played a $\Diamond$-move in η, he did not pick any pointed model of the form $(\mathcal{A}, \infty)$, and

(ii) if Hercules played a $\Box$-move in η, he did not pick any pointed model of the form $(\mathcal{B}, \infty)$.

Proof. The claims are symmetric, so we prove the first. If Hercules picked a frame of the form $(\mathcal{A}, \infty)$, since the Hydra plays greedily, she will also pick at least one pointed model of the form $(\mathcal{B}, \infty)$. By Lemma 7.22, $(\mathcal{A}, \infty) \leftrightarrows (\mathcal{B}, \infty)$, which in view of Lemma 6.6 contradicts the assumption that T is closed. $\Box$

We will not give a full proof of Theorem 7.3.(b), as it proceeds by replacing $\mathbf{A}^n \circ \mathbf{B}^n$ by $\hat{\mathbf{A}}^n \circ \hat{\mathbf{B}}^n$ throughout Section 7.2. Instead, we give a rough outline below.

Proposition 7.25. *For every* $n \geq 1$*, Hercules has no winning strategy of less than* 2^n *moves in the* $(\varphi_n, \hat{\mathbf{A}}^n \cup \hat{\mathbf{B}}^n)$*-MEG.*

Proof sketch. The analogues of Lemmas 7.11 and Lemmas 7.12 for $\hat{\mathbf{A}}^n \circ \hat{\mathbf{B}}^n$ follow directly from the original statements by observing that $S_{\hat{\mathcal{K}}} = S_{\mathcal{K}}$, so that the critical branch is identical. Similarly, an analogue of Lemma 7.13 follows easily from the original if we use Lemma 7.22 to see that, whenever $\boldsymbol{a}' \leftrightarrows \boldsymbol{b}'$, we also have that $\hat{\boldsymbol{a}}' \leftrightarrows \hat{\boldsymbol{b}}'$. The analogue of Lemma 7.15 can then be proven as before, using Lemma 7.24 to rule out situations where Hercules chooses ∞.

The rest of the results leading up to Proposition 7.4 rely on these basic lemmas and thus readily apply to the $(\varphi_n, \mathbf{TC})$-MEG. In particular, Definition 7.16 makes no assumption about the frames appearing in T, hence the sets $\Lambda(i)$ for $i \in [1, 2^n]$ are readily available for the $(\varphi_n, \mathbf{TC})$-MEG, and as before are disjoint and non-empty. We conclude that Hercules has no winning strategy in less than 2^n moves. □

Theorem 7.3(b) readily follows, as does the following stronger version of Proposition 7.4.2:

Proposition 7.26. *For all* $n \geq 1$*, whenever* $\psi \in \mathcal{L}_{\Diamond}$ *is such that* $\varphi_n \equiv \psi$ *on* $\hat{\mathbf{A}}^n \cup \hat{\mathbf{B}}^n$*, it follows that* $|\psi| \geq 2^n$*.*

Remark 7.27. *In fact, in Theorem 7.3(b),* **TC** *can readily be replaced by* $\hat{\mathbf{GL}}$*. While the latter class is not necessarily of independent interest, it is a sub-class of several others studied in the modal logic literature. In particular, any* $\mathcal{A} \in \hat{\mathbf{GL}}$ *has a greatest element and is* Noetherian; *if* $x_0 \; R_{\mathcal{A}} \; x_1 \; R_{\mathcal{A}} \; x_2 \; R_{\mathcal{A}} \; \ldots$*, then there is* $n \in \mathbb{N}$ *such that, for all* $m \geq n$*,* $x_m = x_n$*. Thus,* $\hat{\mathbf{GL}}$ *is a sub-class of the class of transitive, serial, Noetherian frames with a greatest element.*

Note that the property of having a greatest element is quite strong; it implies, for example, that $\mathcal{A}$ *is* confluent, *in the sense that if* $x \; R_{\mathcal{A}} \; y_1$ *and* $x \; R_{\mathcal{A}} \; y_2$*, then there is* $z \in |\mathcal{A}|$ *such that* $y_1 \; R_{\mathcal{A}} \; z$ *and* $y_1 \; R_{\mathcal{A}} \; z$*.*

8 Succinctness in the extended language

Proposition 7.4 holds even if we replace $\mathcal{L}_{\Diamond}$ by the extended language $\mathcal{L}_{\Diamond\forall}^{\Diamond^*}$. In this section, we will first extend these results to $\mathcal{L}_{\Diamond\forall}$ and then to $\mathcal{L}_{\Diamond\forall}^{\Diamond^*}$.

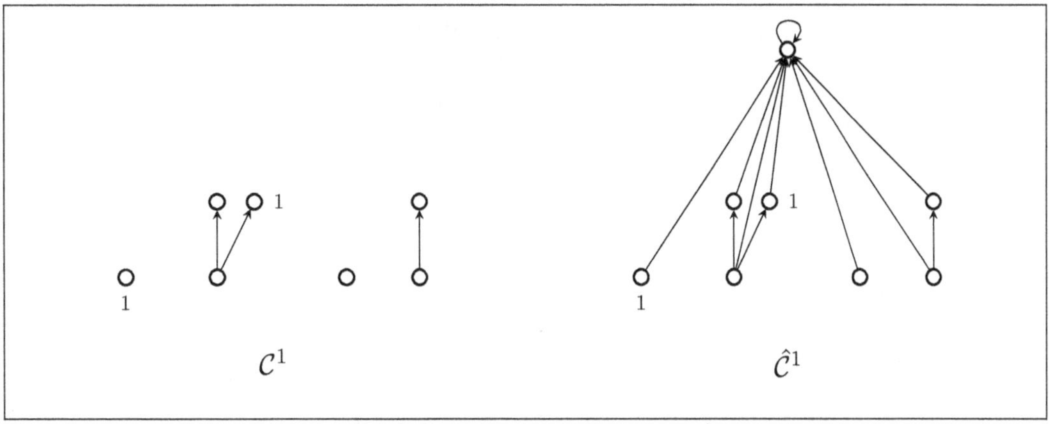

Figure 8: The models $\mathcal{C}^1$ and $\hat{\mathcal{C}}^1$.

8.1 Succinctness with the universal modality

As it turns out, the universal modality is not of much help in expressing $\diamondplus$ succinctly. This is perhaps not surprising, as $\diamondplus$ is essentially a local operator and $\forall$ is global. The disjoint union operation from Section 4.3 will be useful in making this precise.

Definition 8.1. *Let the model $\mathcal{C}^n$ be obtained by taking the disjoint union of all the models used in $\mathbf{A}^n$ and $\mathbf{B}^n$ i.e.,*

$$\mathcal{C}^n = \Big(\coprod_{i=1}^{2^n} \mathcal{A}_i^n\Big) \sqcup \Big(\coprod_{i=1}^{2^n} \mathcal{B}_i^n\Big).$$

The models $\mathcal{C}^n$ will allow us to make the universal modality effectively useless in distinguishing between the pointed models we have constructed.

Proposition 8.2. *For all $n \geq 1$, whenever $\psi \in \mathcal{L}_{\diamond\forall}$ is such that $\varphi_n \equiv \psi$ either on $\mathcal{C}^n$ or on $\hat{\mathcal{C}}^n$, it follows that $|\psi| \geq 2^n$.*

Proof. Given a model $\mathcal{M}$, consider a translation $t_\forall^{\mathcal{M}}: \mathcal{L}_{\diamond\forall} \to \mathcal{L}_\diamond$ that commutes with all Booleans and $\diamond, \Box$, and so that if θ is of one of the forms $\forall\varphi$ or $\exists\varphi$, then $t_\forall^{\mathcal{M}}(\theta) = \top$ if $\mathcal{M} \models \theta$, $t_\forall^{\mathcal{M}}(\theta) = \bot$ otherwise. It is immediately clear that for $w \in |\mathcal{M}|$ and any $\psi \in \mathcal{L}_{\diamond\forall}$, $(\mathcal{M}, w) \models \psi$ if and only if $(\mathcal{M}, w) \models t_\forall^{\mathcal{M}}(\psi)$; moreover, it is obvious that $|t_\forall^{\mathcal{M}}(\psi)| \leq |\psi|$.

Let us consider a point w in one of the models $\mathcal{A}_j^n$ from Definition 7.6. Since $(\mathcal{A}_j^n, w)$ is locally bisimilar to $(\mathcal{C}^n, w)$, by Lemma 4.2 we have that for any $\mathcal{L}_\diamond$-formula ψ, $(\mathcal{C}^n, w) \models \psi$ if and only if $(\mathcal{A}_j^n, w) \models \psi$.

Thus, for any $\mathcal{L}_{\Diamond\forall}$-formula ψ that is equivalent to φ_n on $\mathcal{C}^n$, we have that its translation $t_{\forall}^{\mathcal{C}^n}(\psi)$ is also equivalent to φ_n on $\mathbf{A}^n \cup \mathbf{B}^n$. Note, however, that according to Proposition 7.20, the size of $t_{\forall}^{\mathcal{C}^n}(\psi)$ is at least 2^n. This establishes the proposition for $\mathcal{C}^n$.

For $\hat{\mathcal{C}}^n$ we proceed as above, but work instead with Proposition 7.26. □

8.2 Succinctness with tangle and fixed points

Recall that the translation $t^{\perp}_{\Diamond^*}(\varphi)$ is defined simply by replacing every occurrence of $\Diamond^*\Phi$ by $\perp$ and every occurrence of $\Box^*\Phi$ by $\top$. Since Proposition 7.4 applies to scattered spaces, we can use our work in Section 5.2 to immediately obtain succinctness results relative to the tangled derivative over such spaces.

For metric spaces, however, the behavior of the tangled derivative is less trivial. Fortunately, in the **TC** models we have constructed, its behavior is still rather simple.

Lemma 8.3. *Let $\mathcal{K}$ be a* **GL** *model and $\Phi \subseteq \mathcal{L}^{\Diamond^*}_{\Diamond\forall}$ be finite. Then, for any $w \in |\hat{\mathcal{K}}|$,*

(a) $(\hat{\mathcal{K}}, w) \models \Diamond^*\Phi$ if and only if $(\hat{\mathcal{K}}, \infty) \models \bigwedge \Phi$, and

(b) $(\hat{\mathcal{K}}, w) \models \Box^*\Phi$ if and only if $(\hat{\mathcal{K}}, \infty) \models \bigvee \Phi$.

Proof. Let $\mathcal{K}$ be any **GL** model. By the semantics of $\Diamond^*$, if $w \in |\hat{\mathcal{K}}|$ and $\Phi \subseteq \mathcal{L}^{\Diamond^*}_{\Diamond\forall}$ is finite, then $(\hat{\mathcal{K}}, w) \models \Diamond^*\Phi$ if and only if there is an infinite sequence

$$w = w_0 \, R_{\hat{\mathcal{K}}} \, w_1 \, R_{\hat{\mathcal{K}}} \, w_2 \ldots$$

such that each formula of Φ holds on $(\hat{\mathcal{K}}, w_n)$ for infinitely many values of w_n. However, since $\mathcal{K}$ is a **GL** frame, we must have $w_n = \infty$ for some value of n, which implies that $w_m = \infty$ for all $m \geq n$. From this it readily follows that $(\hat{\mathcal{K}}, \infty) \models \bigwedge \Phi$. Conversely, if $(\hat{\mathcal{K}}, \infty) \models \bigwedge \Phi$, then the sequence $\infty, \infty, \infty, \ldots$ witnesses that $(\hat{\mathcal{K}}, w) \models \Diamond^*\Phi$. It follows that $(\mathcal{M}, w) \models \Diamond^*\Phi$ if and only if $(\mathcal{M}, \infty) \models \bigwedge \Phi$. By similar reasoning, $(\mathcal{M}, w) \models \Box^*\Phi$ if and only if $(\mathcal{M}, \infty) \models \bigvee \Phi$. □

This allows us to define a simple translation from $\mathcal{L}^{\Diamond^*}_{\Diamond\forall}$ to $\mathcal{L}_{\Diamond\forall}$ tailored for our **TC** models.

Definition 8.4. *Fix a* **GL** *model $\mathcal{K}$. We define a translation $t^{\mathcal{K}}_{\Diamond^*} : \mathcal{L}^{\Diamond^*}_{\Diamond\forall} \to \mathcal{L}_{\Diamond\forall}$ by letting $t^{\mathcal{K}}_{\Diamond^*}$ commute with Booleans and all modalities except $\Diamond^*, \Box^*$, in which case we set*

$$t^{\mathcal{K}}_{\Diamond^*}(\Diamond^*\Phi) = \begin{cases} \top & \textit{if } (\hat{\mathcal{K}}, \infty) \models \bigwedge \Phi, \\ \bot & \textit{otherwise;} \end{cases} \qquad t^{\mathcal{K}}_{\Diamond^*}(\Box^*\Phi) = \begin{cases} \top & \textit{if } (\hat{\mathcal{K}}, \infty) \models \bigvee \Phi, \\ \bot & \textit{otherwise.} \end{cases}$$

Lemma 8.5. *Let $\mathcal{K}$ be a* **GL** *model and $\varphi \in \mathcal{L}^{\Diamond^*}_{\Diamond\forall}$ be finite. Then, $t^{\mathcal{K}}_{\Diamond^*}(\varphi) \equiv \varphi$ over $\hat{\mathcal{K}}$.*

Proof. Immediate from Lemma 8.3 using a routine induction on φ. □

We are now ready to prove the full version of our first main result:

Theorem 8.6. *Let* **C** *be a class of convergence spaces that contains either*

1. *all finite* **GL** *frames,*
2. *all finite* **TC** *frames,*
3. *all ordinals $\Lambda < \omega^\omega$, or*
4. *any crowded metric space $\mathcal{X}$.*

Then there exist arbitrarily large $\varphi \in \mathcal{L}_{\oplus}$ such that, whenever $\psi \in \mathcal{L}^{\Diamond^*}_{\Diamond\forall}$ is equivalent to φ over **C***, it follows that $|\psi| \geq 2^{\frac{|\varphi|}{3}}$.*

Proof. First assume that **C** contains all finite **GL** frames. Fix $n \geq 1$, and assume that $\psi \in \mathcal{L}^{\Diamond^*}_{\Diamond\forall}$ is equivalent to φ_n over **C**. Then, by Corollary 5.10, $t^{\bot}_{\Diamond^*}(\psi) \in \mathcal{L}_{\Diamond\forall}$ is equivalent to ψ, and hence to φ_n, over **C**, and in particular, over $\mathcal{C}^n$. By Proposition 8.2, it follows that $|t^{\bot}_{\Diamond^*}(\psi)| \geq 2^n$, and hence $|\psi| \geq 2^n$ as well. This establishes item 1 and, in view of Corollary 5.13, item 3.

If **C** contains the class of all finite **TC** frames, we proceed as above, but instead use the translation $t^{\mathcal{C}^n}_{\Diamond^*}$, so that $t^{\mathcal{C}^n}_{\Diamond^*}(\psi) \in \mathcal{L}_{\Diamond\forall}$ is equivalent to $\psi \in \mathcal{L}^{\Diamond^*}_{\Diamond\forall}$, and hence to φ_n, over $\hat{\mathcal{C}}^n$. Finally, we use Corollary 5.5 to lift this result to **C** containing any crowded metric space $\mathcal{X}$. □

Given the fact that $\mathcal{L}^{\Diamond^*}_{\Diamond}$ is equally expressive as $\mathcal{L}^{\mu}_{\Diamond}$, it is natural to ask which is more succinct. Note that $\mathcal{L}^{\Diamond^*}_{\Diamond}$ cannot be exponentially more succinct than $\mathcal{L}^{\mu}_{\Diamond}$, as the translation $t^{\mu}_{\Diamond^*}$ is polynomial. On the other hand, we do have that the μ-calculus is more succinct than the tangled language:

Theorem 8.7. *Let* **C** *contain either*

1. *all finite* **GL** *frames,*
2. *all finite* **TC** *frames,*
3. *all ordinals $\Lambda < \omega^\omega$, or*
4. *any crowded metric space $\mathcal{X}$.*

Then there exist arbitrarily large $\theta \in \mathcal{L}^{\mu}_{\Diamond}$ such that, whenever $\psi \in \mathcal{L}^{\Diamond^*}_{\Diamond\forall}$ is equivalent to θ over $\mathbf{C}$, *it follows that $|\psi| > 2^{\frac{|\theta|}{12}}$.*

Proof. By Lemma 3.2, for all $n \in \mathbb{N}$, $\varphi_n \equiv t^{\mu}_{\oplus}(\varphi_n)$ and $|t^{\mu}_{\oplus}(\varphi_n)| \leq 4|\varphi_n|$. Hence, the sequence $(t^{\mu}_{\oplus}(\varphi_n))_{n\in\mathbb{N}}$ witnesses that $\mathcal{L}^{\mu}_{\Diamond}$ is exponentially more succinct than $\mathcal{L}^{\Diamond^*}_{\Diamond\forall}$. □

Remark 8.8. *In view of Remark 7.27, the class* **TC** *can be replaced by* $\hat{\mathbf{GL}}$ *in both Theorems 8.6 and 8.7.*

9 Concluding remarks

There are several criteria to take into account when choosing the 'right' modal logic for spatial reasoning. It has long been known that the limit-point operator leads to a more expressive language than the closure operator does, making the former seem like a better choice of primitive symbol. However, our main results show that one incurs in losses with respect to formula-size, and since the blow-up is exponential, this could lead to situations where e.g. formally proving a theorem expressed with the closure operator is feasible, but treating its limit-operator equivalent is not. Similarly, the results of Dawar and Otto [9] make the tangled limit operator seem like an appealing alternative to the spatial μ-calculus, given its simpler syntax and more transparent semantics. Unfortunately, the price to pay is also an exponential blow-up.

We believe that the takeaway is that different modal logics may be suitable for different applications, and hope that the work presented here can be instrumental in clarifying the advantages and disadvantages of each option. Moreover, there are many related questions that remain open and could give us a more complete picture of the relation between such languages.

For example, one modality that also captures interesting spatial properties is the 'difference' modality, where $\langle\neq\rangle\varphi$ holds at w if there is $v \neq w$ satisfying φ. This modality has been studied in a topological setting by Kudinov [26], and succinctness between languages such as $\mathcal{L}_{\Diamond\forall}$ and $\mathcal{L}_{\oplus\langle\neq\rangle}$ remains largely unexplored. Even closer to the present work is the *tangled closure* operator, $\oplus^*$, defined analogously to $\Diamond^*$, but instead using the closure operation. The techniques we have developed here do not settle whether $\mathcal{L}^{\mu}_{\oplus}$ is exponentially more succinct than $\mathcal{L}^{\oplus^*}_{\oplus}$.

There are also possible refinements of our results. Our construction uses infinitely many variables, and it is unclear if $\mathcal{L}_{\oplus}$ is still exponentially more succinct than $\mathcal{L}_{\Diamond}$ when restricted to a finite number of variables. Finally, note that Theorem 8.6 is sharp in the sense that an exponential translation is available, but this is not so clear for Theorem 8.7, in part because an explicit translation is not given by Dawar and Otto [9]. Sharp upper and lower bounds remain to be found.

Acknowledgements

We are grateful to the anonymous reviewers whose detailed comments and suggestions helped greatly improve the presentation of our results.

References

[1] M. Abashidze. Ordinal completeness of the Gödel-Löb modal system. *Intensional Logics and the Logical Structure of Theories*, pages 49–73, 1985. in Russian.

[2] M. Adler and N. Immerman. An $n!$ lower bound on formula size. *ACM Transactions on Computational Logic*, 4(3):296–314, 2003.

[3] J. P. Aguilera and D. Fernández-Duque. Strong completeness of provability logic for ordinal spaces. *Journal of Symbolic Logic*, 82(2):608–628, 2017.

[4] G. Bezhanishvili, L. Esakia, and D. Gabelaia. Some results on modal axiomatization and definability for topological spaces. *Studia Logica*, 81(3):325–355, 2005.

[5] G. Bezhanishvili and M. Gehrke. Completeness of S4 with respect to the real line: revisited. *Annals of Pure and Applied Logic*, 131(1-3):287–301, 2005.

[6] A. Blass. Infinitary combinatorics and modal logic. *Journal of Symbolic Logic*, 55(2):761–778, 1990.

[7] G. S. Boolos. *The Logic of Provability.* Cambridge University Press, 1993.

[8] A. Chagrov and M. Zakharyaschev. *Modal Logic*, volume 35 of *Oxford logic guides*. Oxford University Press, 1997.

[9] A. Dawar and M. Otto. Modal characterisation theorems over special classes of frames. *Annals of Pure and Applied Logic*, 161:1–42, 2009.

[10] K. Eickmeyer, M. Elberfeld, and F. Harwath. Expressivity and succinctness of order-invariant logics on depth-bounded structures. In *Proceedings of MFCS 2014*, pages 256–266, 2014.

[11] K. Etessami, M. Vardi, and T. Wilke. First-order logic with two variables and unary temporal logic. In *Proceedings of the Twelfth Annual IEEE Symposium on Logic in Computer Science (LICS)*, pages 228–236, 1997.

[12] D. Fernández-Duque. Tangled modal logic for spatial reasoning. In *Proceedings of IJCAI*, pages 857–862, 2011.

[13] D. Fernández-Duque. Tangled modal logic for topological dynamics. *Annals of Pure and Applied Logic*, 163(4):467–481, 2012.

[14] D. Fernández-Duque. Non-finite axiomatizability of dynamic topological logic. *ACM Transactions on Computational Logic*, 15(1):4:1–4:18, 2014.

[15] S. Figueira and D. Gorín. On the size of shortest modal descriptions. *Advances in Modal Logic*, 8:114–132, 2010.

[16] T. French, W. van der Hoek, P. Iliev, and B. Kooi. Succinctness of epistemic languages. In T. Walsh, editor, *Proceedings of IJCAI*, pages 881–886, 2011.

[17] T. French, W. van der Hoek, P. Iliev, and B. Kooi. On the succinctness of some modal logics. *Artificial Intelligence*, 197:56–85, 2013.

[18] G. Gogic, C. Papadimitriou, B. Selman, and H. Kautz. The comparative linguistics of knowledge representation. In *Proceedings of IJCAI*, pages 862–869, 1995.

[19] R. Goldblatt and I. Hodkinson. The tangled derivative logic of the real line and zero-dimensional space. In *Advances in Modal Logic*, volume 11, pages 342–361, 2016.

[20] R. Goldblatt and I. Hodkinson. The finite model property for logics with the tangle modality. *Studia Logica*, 2017.

[21] R. Goldblatt and I. Hodkinson. Spatial logic of tangled closure operators and modal mu-calculus. *Annals of Pure and Applied Logic*, 168(5):1032 – 1090, 2017.

[22] M. Grohe and N. Schweikardt. The succinctness of first-order logic on linear orders. *Logical Methods in Computer Science*, 1:1–25, 2005.

[23] L. Hella and M. Vilander. The succinctness of first-order logic over modal logic via a formula size game. In *Advances in Modal Logic*, volume 11, pages 401–419, 2016.

[24] K. Hrbacek and T.J. Jech. *Introduction to set theory*. Monographs and textbooks in pure and applied mathematics. M. Dekker, 1984.

[25] P. Kremer. Strong completeness of S4 for any dense-in-itself metric space. *Reveiw of Symbolic Logic*, 6(3):545–570, 2013.

[26] A. Kudinov. Topological modal logics with difference modality. In *Advances in Modal Logic*, volume 6, pages 319–332, 2006.

[27] A. Kudinov and V. Shehtman. Derivational modal logics with the difference modality. In Guram Bezhanishvili, editor, *Leo Esakia on Duality in Modal and Intuitionistic Logics*, pages 291–334, Dordrecht, 2014. Springer Netherlands.

[28] K. Kuratowski. *Topology*. Academic Press, first edition, 1966.

[29] J. Lucero-Bryan. The d-logic of the real line. *Journal of Logic and Computation*, 23(1):121–156, February 2013.

[30] J.C.C. McKinsey and A. Tarski. The algebra of topology. *Annals of Mathematics*, 2:141–191, 1944.

[31] G. Mints and T. Zhang. A proof of topological completeness for S4 in $(0, 1)$. *Annals of Pure and Applied Logic*, 133(1-3):231–245, 2005.

[32] J. R. Munkres. *Topology*. Prentice Hall, 2nd ed edition, 2000.

[33] E. Pacuit. *Neighborhood Semantics for Modal Logic*. Springer, 2017.

[34] H. Rasiowa and R. Sikorski. *The mathematics of metamathematics*. Państwowe Wydawnictwo Naukowe, Warsaw, 1963.

[35] A. Razborov. Applications of matrix methods to the theory of lower bounds in computational complexity. *Combinatorica*, 10(1):81–93, 1990.

[36] V. Shehtman. Derived sets in Euclidean spaces and modal logic. *University of Amsterdam Technical Report*, X-1990-05, 1990.

[37] V. Shehtman. 'Everywhere' and 'here'. *Journal of Applied Non-Classical Logics*, 9(2-3):369–379, 1999.

[38] R. M. Solovay. Provability interpretations of modal logic. *Israel Journal of Mathematics*, 28:33–71, 1976.

[39] A. Tarski. A lattice-theoretical fixpoint theorem and its applications. *Pacific J. Math.*, 5(2):285–309, 1955.

[40] W. van der Hoek, P. Iliev, and B. Kooi. On the relative succinctness of modal logics with union, intersection and quantification. In *Proceedings of AAMAS*, pages 341–348, 2014.

[41] T. Wilke. CTL^+ is exponentially more succinct than CTL. In *Proceedings of Nineteenth Conference on Foundations of Software Technology and Theoretical Computer Science (FSTTCS)*, pages 110–121, 1999.

Received 12 August 2017

Formal Analysis of Discrete-Time Systems using z-Transform

Umair Siddique
Department of Electrical and Computer Engineering, Concordia University, Montreal, QC, Canada
muh_sidd@ece.concordia.ca

Mohamed Yousri Mahmoud
Department of Electrical and Computer Engineering, Concordia University, Montreal, QC, Canada
mo_solim@encs.concordia.ca

Sofiène Tahar
Department of Electrical and Computer Engineering, Concordia University, Montreal, QC, Canada
tahar@ece.concordia.ca

Abstract

The computer implementation of a majority of engineering and physical systems requires the discretization of continuous parameters (e.g., time, temperature, voltage, etc.). Such systems are then called discrete-time systems and their dynamics can be described by difference or recurrence equations. Recently, there is an increasing interest in applying formal methods in the domain of cyber-physical systems to identify subtle but critical design bugs, which can lead to critical failures and monetary loss. In this paper, we propose to formally reason about discrete-time aspects of cyber-physical systems using the z-Transform, which is a mathematical tool to transform a time-domain model to a corresponding complex-frequency domain model. In particular, we present the HOL Light formalization of the z-Transform and difference equations along with some important properties such as linearity, time-delay and complex translation. An interesting part of our work is the formal proof of the uniqueness of the z-Transform. Indeed, the uniqueness of the z-Transform plays a vital role in reliably deducing important properties of complex systems. We apply our work to formally analyze a switched-capacitor interleaved DC-DC voltage doubler and an infinite impulse response (IIR) filter, which are important components of a wide class of power electronics, control and signal processing systems.

Vol. 5 No. 4 2018
Journal of Applied Logics — IFCoLog Journal of Logics and their Applications

1 Introduction

We observe many continuous-time natural phenomena in our every day life, for instance the speed of a car, the temperature of a city and heart-beat are time varying quantities. Even though continuous-time quantities permeate in nature, we also observe many discrete-time quantities, e.g., maximum and minimum temperature in a city, average speed of traffic vehicles and a stock market index. It is therefore indispensable to design engineering systems which can detect and process these phenomena to achieve different functionalities. However, the continuous-time quantities cannot be processed directly using digital computing machines, which are suitable to deal with the discrete-time quantities. In practice, a continuous-time quantity is converted to a corresponding sampled version which coincides with the original quantity at some instant in time [6]. For example, a continuous-time signal can be sampled into a sequence of numbers where each number is separated from the next in time by a sampling period of T seconds as shown in Figure 1.

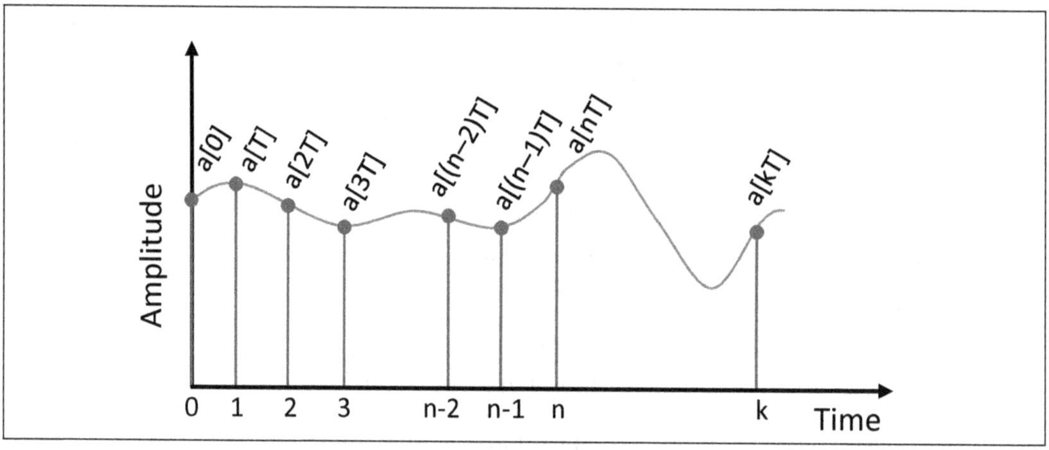

Figure 1: Sampling of a Continuous Signal

In general, the dynamics of engineering and physical systems are characterized by differential equations [33] and difference equations [7] in case of continuous-time and discrete-time, respectively. The complexity of these equations varies depending upon the corresponding system architecture (distributed, cascaded, hybrid etc.), the nature of input signals and the physical constraints. Transformation analysis is one of the most efficient techniques to mathematically analyze such complex systems. The main objective of transform method is to reduce complicated system models (i.e., differential or difference equations) into algebraic equations. The z-Transform [21] provides a mechanism to map discrete-time signals over the complex plane also called z-domain. This transform is a powerful tool to solve linear difference equations (LDE) by transforming them into algebraic operations in z-domain. Moreover, the z-domain representation of LDEs is also used for the transfer function analysis of corresponding systems. Due to these distinctive features, the z-Transform is one of the main core techniques available in physical and engineering system analysis software tools (e.g., MATLAB [20], Mathematica[19]) and is widely used in the design and analysis of signal processing filters [21], electronic circuits [7], control systems [8], photonic devices [5] and queueing networks [1].

The main idea of the z-Transform can be traced back to Laplace, but it was formally introduced by W. Hurewicz (1947) to solve linear constant coefficient difference equations [15]. Mathematically, the z-Transform can be defined as a function series which transforms a discrete time signal $f[n]$ to a function of a complex variable z, as follows:

$$X(z) = \sum_{n=0}^{\infty} f[n]z^{-n} \tag{1}$$

where $f[n]$ is a complex-valued function ($f : \mathbb{N} \rightarrow \mathbb{C}$) and the series is defined for those $z \in \mathbb{C}$ for which the series is convergent.

The first step in analyzing a difference equation (e.g., $x_{n+1} = kx_n(1 - x_n)$) using the z-Transform is to apply the z-Transform on both sides of a given equation. Next, the corresponding z-domain equation is simplified using various properties of the z-Transform, such as linearity, scaling and differentiation. The main task is to either solve the difference equation or to find a transfer function which relates the input and output of the corresponding system. Once the transfer function is obtained, it can be used to analyze some important aspects such as stability, frequency response and design optimization to reduce the number of corresponding circuit elements such as multipliers and shift registers.

Traditionally, the analysis of linear systems based on the z-Transform has been done using numerical computations and symbolic techniques [20, 19]. Both of these approaches, including paper-and-pencil proofs [21] have some known limitations like incompleteness, numerical errors and human-error proneness. In recent years, theorem proving has been actively used for both the formalization of mathematics (e.g., [11, 9]) and the analysis of physical systems (e.g., [30, 29]). For the latter case, the main task is to identify and formalize the underlying mathematical theories. In practice, four fundamental transformation techniques (i.e., the Laplace Transform (LT), the z-Transform (ZT), the Fourier Transform (FT), and the Discrete Fourier Transform (DFT)) are used in the design and development of linear systems. Interestingly, the Fourier transform and the Discrete Fourier transform can be derived from the Laplace Transform and the z-Transform, respectively. The formalization of the Laplace Transform and the Fourier Transform have been reported in [32] and [22] using the multivariate analysis libraries of HOL Light [12], with an ultimate goal of reasoning about differential equations and transfer functions of continuous systems. However, the formal proof of both the inverse Laplace Transform and the inverse Fourier Transform have not been provided in [32] and [22], which is necessary to reason about transformation from s-domain and ω-domain (where s and ω are LT and FT domain parameters, respectively) to the time-domain. The uniqueness and inverse of the z-Transform can be used to overcome this limitation by using the well-known Bilinear-Transformation of the z-domain and the s-domain [21]. The main relation amongst these four transformations is outlined in Figure 2.

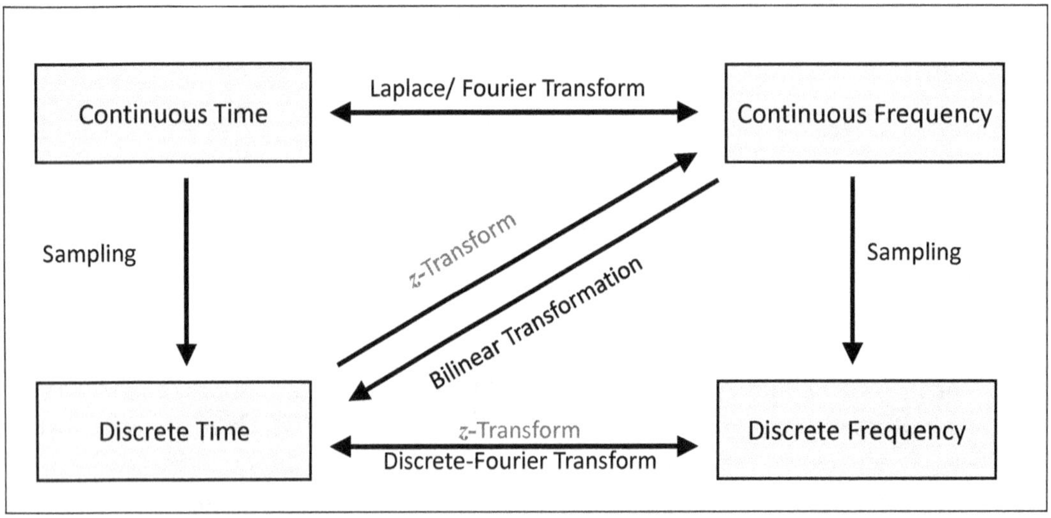

Figure 2: Discrete and Continuous Transformation Analysis

Nowadays, discrete-time linear systems are widely used in safety and mission critical domains (e.g., digital control of avionics systems and biomedical devices). We believe that there is a dire need for an infrastructure which provides the basis for the formal analysis of discrete-time systems within the sound core of a theorem prover. In this paper, we propose a formal analysis approach for the z-Transform based system models using a higher-order logic (HOL) theorem prover. The main idea is to leverage upon the high expressiveness of HOL to formalize Equation (1) and use it to verify classical properties of the z-Transform within a theorem prover. These foundations can be built upon to reason about the analytical solutions of difference equations or transfer functions. In [28], we presented the preliminary formalization of the z-Transform and its associated region of convergence (ROC). In this paper, however, we widen the scope by adding more interesting properties such as complex conjugation and initial value theorem of the z-Transform. We also provide the formally verified expressions for the z-Transform of commonly used mathematical functions (e.g., $\exp(x)$, $\sin(x)$ and $\cos(x)$). We then present the formalization of generic linear constant coefficient difference equations (LCCDE) along with the formal verification of corresponding z-Transform expressions by utilizing the key properties such as linearity of the z-Transform and ROC. A central part of the reported work is the formal proof of the uniqueness and inverse of the z-Transform, for which we also formalize the notion of an exterior region of a circle and its relation to ROC of the z-Transform. In order to demonstrate the practical effectiveness of the reported work, we present the formal analysis of a switched capacitor DC-DC power converter and an infinite impulse response (IIR) digital signal processing filter. The formalization reported in this paper has been developed in the latest version of the HOL Light theorem prover due to its rich multivariate analysis libraries [12]. The source code of our formalization is available for download [23] and can be utilized by other researchers and engineers for further developments and the analysis of more practical systems.

The rest of the paper is organized as follows: Section 2 describes some fundamentals of multivariate analysis libraries of the HOL Light theorem prover. Sections 3 and 4 present our HOL Light formalization of the z-Transform and the verification of its properties, respectively. Section 5 presents the formalization of difference equations and transfer functions. We describe the formal proof of the uniqueness of the z-Transform in Section 6. In Section 7, we present the analysis of a power-electronic DC-DC converter and IIR filter. Finally, Section 8 concludes the paper and highlights some future directions.

2 Preliminaries

In this section, we provide a brief introduction to the HOL Light formalization of some core concepts such as vector summation, summability, complex differentiation and infinite summation [12]. Our main intent is to introduce the basic definitions and notations that are used in the rest of the paper.

In the formalization of multivariate theory, an N-dimensional vector is represented as an $\mathbb{R}^N$ column matrix with individual elements as real numbers. All of the vector operations are then treated as matrix manipulations. Similarly, instead of defining a new type, complex numbers ($\mathbb{C}$) can be represented as $\mathbb{R}^2$. Most of the theorems available in multivariate libraries of HOL Light are verified for arbitrary functions with a flexible data-type of ($\mathbb{R}^M \rightarrow \mathbb{R}^N$). The injection from natural numbers to complex numbers can be represented by $\& : \mathbb{N} \rightarrow \mathbb{R}$. Similarly, the injection from real to complex numbers is done by $\texttt{Cx} : \mathbb{R} \rightarrow \mathbb{C}$. The real and imaginary parts of a complex number are represented by `Re` and `Im` both with type $\mathbb{C} \rightarrow \mathbb{R}$. The unary negation of x is represented as $-x$, where x can be real or a complex number.

The generalized summation over arbitrary functions is defined as follows:

Definition 1 (Vector Summation).

$\vdash_{def}$ `∀s f. vsum s f = (lambda i. sum s (λx. f x$i))`

where `vsum` takes two parameters $\texttt{s} : \texttt{A} \rightarrow \texttt{bool}$ which specifies the set over which the summation occurs and an arbitrary function $\texttt{f} : (\texttt{A} \rightarrow \mathbb{R}^{\texttt{N}})$. The function `sum` is a finite summation over real numbers and accepts $\texttt{f} : (\texttt{A} \rightarrow \mathbb{R}^{\texttt{N}})$. For example, $\sum_{i=0}^{K} f(i)$ can be represented as `vsum (0..K) f`.

The traditional mathematical expression $\sum_{i=0}^{\infty} f(i) = L$ is defined in HOL Light as follows:

Definition 2 (Sums).

$\vdash_{def}$ `∀s f l. (f sums l) s ⇔`
`((λn. vsum (s INTER (0..n)) f) ⟶ l) sequentially`

where the types of the parameters are: $(\texttt{s} : \mathbb{N} \rightarrow \texttt{bool})$, $(\texttt{f} : \mathbb{N} \rightarrow \mathbb{R}^{\texttt{N}})$ and $(\texttt{L} : \mathbb{R}^{\texttt{N}})$.

We present the definition of the summability of a function $(\texttt{f} : \mathbb{N} \rightarrow \mathbb{R}^{\texttt{N}})$, which indeed represents that there exist some $(\texttt{L} : \mathbb{R}^{\texttt{N}})$ such that $\sum_{i=0}^{\infty} f(i) = L$.

Definition 3 (Summability).

$\vdash_{def}$ `∀f s. summable s f ⇔ (∃l. (f sums l) s)`

The limit of an arbitrary function can be defined as follows:

Definition 4 (Limit).

$\vdash_{def}$ `∀f net. lim net f = (εl. (f ⟶ l) net)`

where the function `lim` is defined using the Hilbert choice operator ε in the functional form. It accepts a `net` with elements of arbitrary data-type `A` and a function $(\texttt{f} : \texttt{A} \rightarrow \mathbb{R}^{\texttt{N}})$, and returns $(\texttt{L} : \mathbb{R}^{\texttt{N}})$ the value to which `f` converges at the given `net`. In Definition 2, `sequentially` represents a sequential net which describes the sequential evolution of a function, i.e., $f(i), f(i+1), f(i+2), \ldots$, etc. This is typically used in the definition of an infinite summation. Note that `nets` are defined as a bijective type in which domain is the set of two-parameter boolean functions, where we use the function `mk_net` to construct a net. The sequential nets are defined as `mk_net` λ`m n. m` $\geq$ `n`. According to this definition, we notice that the number `a` that satisfies the property $\forall$`n. (n` $\geq$ `a)`, represents infinity. The continuous counterpart of the sequential net is `at_infinity`, which is defined as `mk_net` λx y. norm(x) $\geq$ norm(y). This is a generalized definition valid for any Euclidian space $\mathbb{R}^{\mathbb{N}}$. In case of real numbers, this simply reduces to `mk_net` λx y. x $\geq$ y. The concept *tends to* ($\longrightarrow$) is formally defined as follows:

Definition 5.

$\vdash_{def}$ ∀f l net. (f $\longrightarrow$ l) net ⇔
(∀e. &0 < e ⇒ eventually(λx.dist (f x,l) < e) net)

We next present the definition of an infinite summation which is one of the most fundamental requirement in our development.

Definition 6 (Infinite Summation).

$\vdash_{def}$ ∀f s. infsum s f = (εl. (f sums l) s)

where function `infsum` is defined using the Hilbert choice operator ε in the functional form. It accepts a parameter (s : num $\rightarrow$ bool) which specifies the starting point and a function (f : $\mathbb{N} \rightarrow \mathbb{R}^{\mathbb{N}}$), and returns (L : $\mathbb{R}^{\mathbb{N}}$) , i.e., the value at which infinite summation of f converges from the given s.

In some situations, it is very useful to specify infinite summation as a limit of finite summation (`vsum`). We proved this equivalence in the following theorem:

Theorem 1 (Infinite Summation in Terms of Sequential Limit).

$\vdash$ ∀s f. infsum s f = lim sequentially(λk.vsum (s INTER (0..k)) f)

The differentiability of complex-valued functions is quite important in the development of the z-Transform, since it is the key element of proving uniqueness of the z-Transform. In HOL Light, a complex derivative is defined using the vector derivative as follows:

Definition 7 (Vector Derivative).

$\vdash_{def}$ ∀f f′ net. (f has_complex_derivative f′) net ⇔
(f has_derivative (λx. f′ * x)) net

where a vector derivative (`has_derivative`) is defined as follows:

Definition 8 (Vector Derivative).

$\vdash_{def}$ ∀f f′ net. (f has_derivative f′) net ⇔
linear f′ ∧ ((λy. inv (norm (y − netlimit net)) %
(f y − (f (netlimit net) +
f′ (y − netlimit net)))) $\longrightarrow$ vec 0) net

where `netlimit` of a `net` returns the supremum of the `net`.

The definition of a complex derivative can also be described in a functional form as follows:

Definition 9 (Complex Differentiation).

```
⊢def ∀f x. complex_derivative f x =
          (εf'. (f has_complex_derivative f') (at x))
```

This definition can further be generalized to formalize the concept of higher-order complex derivatives as described in the following definition:

Definition 10 (Higher-Order Complex Derivative).

```
⊢def ∀f. higher_complex_derivative 0 f = f ∧
       (∀n. higher_complex_derivative (SUC n) f =
        complex_derivative (higher_complex_derivative n f))
```

Another important concept in complex analysis is holomorphic functions which are differentiable in the neighbourhood of every point in their domain. The formal definition of holomorphic functions in HOL Light is given as follows:

Definition 11 (Holomorphic Function).

```
⊢def ∀f s. f holomorphic_on s ⇔
           (∀x. x IN s ⇒
           (∃f'. (f has_complex_derivative f') (at x within s)))
```

3 Formalization of z-Transform

The unilateral z-Transform [16] of a discrete time function $f[n]$ can be defined as follows:

$$F(z) = \sum_{n=0}^{\infty} f[n]z^{-n} \tag{2}$$

where f is a function from $\mathbb{N} \to \mathbb{C}$ and z is a complex variable. Here, the definition that we consider has limits of summation from $n = 0$ to ∞. On the other hand, one can consider these limits from $n = -\infty$ to ∞ and such a version of the z-Transform is called two-sided or bilateral. This generalization comes at the cost of some complications such as non-uniqueness, which limits its practicality in engineering systems analysis. On the other hand, unilateral transform can only be applied to *causal* functions, i.e., $f[n] = 0$ for $\forall n. n < 0$. In practice, unilateral z-Transform is sufficient to analyze most of the engineering systems because their designs involve only causal signals [31]. For similar reasons, the authors of [32] formalized the unilateral Laplace transform rather than the bilateral version.

An essential issue of the z-Transform of $f[n]$ is whether the $F(z)$ even exists, and under what conditions it exists. It is clear from Equation (2) that the z-Transform of a function is an infinite series for each z in the complex plane or z-domain. It is important to distinguish the values of z for which the infinite series is convergent and the set of all those values is called the *region of convergence* (ROC). In mathematics and digital signal processing literature, different definitions of the ROC are considered. For example, one way is to express z in the polar form ($z = re^{j\omega}$) and then the ROC for $F(z)$ includes only those values of r for which the sequence $f[n]r^{-n}$ is absolutely summable. Unfortunately, to the best of our knowledge, this claim (i.e., absolute summability, e.g., [21]) is incorrect for certain functions, for example, $f[n] = \frac{1}{n}u[n-1]$ for which certain values of r result in convergent infinite series, but $x[n]r^{-n}$ is not absolutely summable.

Now, we have two distinct choices for defining the ROC: (1) values of z for which $F(z)$ is finite (or summable) and (2) values of z for which $x[n]z^{-n}$ is absolutely summable. Most of textbooks are not rigorous about the choice of the ROC and both of these definitions are widely used in the analysis of engineering systems. In this paper, we use the first definition of the ROC, which we can define mathematically as follows:

$$ROC = \{z \in \mathbb{C} : \exists k. \sum_{n=0}^{\infty} f[n]z^{-n} = k\} \tag{3}$$

In the above discussion, we mainly highlighted some arbitrary choices of using the definition of the z-Transform and its associated ROC. We formalize the z-Transform function (Equation 2) in HOL Light, as follows:

Definition 12 (z-Transform).

```
⊢def ∀f z. z_transform f z = infsum (from 0) (λn. f n / z pow n)
```

where the `z_transform` function accepts two parameters: a function $\mathtt{f} : \mathbb{N} \to \mathbb{C}$ and a complex variable $\mathtt{z} : \mathbb{C}$. It returns a complex number which represents the z-Transform of `f` according to Equation (2).

We formalize the ROC of the z-Transform as follows:

Definition 13 (Region of Convergence).

```
⊢def ∀f. ROC f = {z | ¬(z = Cx (&0)) ∧
                      summable (from 0) (λn. f n / z pow n)}
```

here, `ROC` accepts a function $\mathtt{f} : \mathbb{N} \to \mathbb{C}$ and returns a set of non-zero values of variable z for which the z-Transform of `f` exists. In order to compute the z-Transform, it is mandatory to specify the associated ROC. We prove two basic properties of ROC which describe the linearity and scaling of the ROC, as follows:

Theorem 2 (ROC Linear Combination).

```
⊢ ∀z a b f g.  z IN ROC f ∧ z IN ROC g  ⇒
               z IN ROC (λn. a * f n) INTER ROC (λn. b * g n)
```

Theorem 3 (ROC Scaling).

```
⊢ ∀z a f. z IN ROC f ⇒  z IN ROC (λn. f n / a)
```

Theorem 2 describes that if z belongs to ROC f and ROC g then it also belongs to the intersection of both ROCs even though the functions f and g are scaled by complex parameters a and b, respectively. Similarly, Theorem 3 shows the scaling with respect to complex division by a complex number a.

4 Main Properties of the z-Transform

In this section, we use Definitions 12 and 13 to formally verify some of the classical properties of the z-Transform in HOL Light. The verification of these properties plays an important role in reducing the time required to analyze practical applications, as described later in Section 7.

4.1 Linearity of the z-Transform

The linearity of the z-Transform is a very useful property while handling systems composed of subsystems with different scaling inputs. Mathematically, it can be defined as:

If $\mathcal{Z}(f[n]) = F(z)$ and $\mathcal{Z}(g[n]) = G(z)$ then the following holds:

$$\mathcal{Z}(\alpha * f[n] \pm \beta * g[n]) = \alpha * F(z) \pm \beta * G(z) \tag{4}$$

The z-Transform of a linear combination of sequences is the same linear combination of the z-Transform of the individual sequences. We verify this property as the following theorem:

Theorem 4 (Linearity of z-Transform).

```
⊢ ∀f g z a b. z IN ROC f ∧ z IN  ROC g ⇒
              z_transform (λx. a * f x + b * g x) z =
              a * z_transform f z + b * z_transform g z
```

where a : $\mathbb{C}$ and b : $\mathbb{C}$ are arbitrary constants. The proof of this theorem is based on the linearity of the infinite summation and Theorem 2.

4.2 Shifting Properties

The shifting properties of the z-Transform are mostly used in the analysis of digital systems and in particular in solving difference equations. In fact, there are two kinds of possible shifts: left shift ($f[n+m]$) or time advance and right shift ($f[n-m]$) or time delay. The main idea is to express the transform of the shifted signal (($f[n+m]$) or ($f[n-m]$)) in terms of its z-Transform ($F(z)$).

Left Shift of a Sequence: If $\mathcal{Z}(f[n])\ z = F(z)$ and k is a positive integer, then the left shift of a sequence can be described as follows:

$$\mathcal{Z}(f[n+k])\ z = z^k(F(z) - \sum_{n=0}^{k-1} f[n]z^{-n}) \tag{5}$$

We verify this theorem as follows:

Theorem 5 (Left Shift or Time Advance).

```
⊢ ∀f z k. z IN ROC f ∧ 0 < k ⇒
          z_transform (λn. f (n + k)) z =
          z pow k * (z_transform f z −
                     vsum (0..k − 1) (λn. f n / z pow n))
```

The verification of this theorem mainly involves properties of complex numbers, summability of shifted functions and splitting an infinite summation into two parts as given by the following lemma:

Lemma 1 (Infsum Splitting).

```
⊢ ∀f n m. summable (from m) f ∧ 0 < n ∧ m ≤ n ⇒
         infsum (from m) f = vsum (m..n−1) f + infsum (from n) f
```

Right Shift of a Sequence: If $\mathcal{Z}(f[n])\ z = F(z)$, and assuming $f(-n) = 0,\ \ \forall n = 1, 2, ..., m$, then the right shift or time delay of a sequence can be described as follows:

$$\mathcal{Z}(f[n-m])\ z = z^{-m}F(z) \tag{6}$$

We formally verify the above property as the following theorem:

Theorem 6 (Right Shift or Time Delay).

```
⊢ ∀f z m. z IN ROC f ∧ is_causal f ⇒
         z_transform (λn. f (n−m)) z = z_transform f z / z pow m
```

Here, `is_causal` defines the causality of the function `f` in a relational form to ensure that $\texttt{f(n} - \texttt{m)} = 0$, $\forall \texttt{m.n} < \texttt{m}$. The proof of this theorem also involves properties of complex numbers along with the following two lemmas:

Lemma 2 (Series Negative Offset).

```
⊢ ∀f k l. (f sums l) (from 0) ⇒  ((λn. f (n−k)) sums l) (from k)
```

Lemma 3 (Infinite Summation Negative Offset).

```
⊢ ∀f k. summable (from 0) f ⇒
        infsum (from 0) (λn. if k ≤ n then f (n−k) else vec 0) =
        infsum (from 0) f
```

As a direct application of the above results, we verify another important property called first-difference (which represents the difference between two consecutive samples of a signal), as follows:

Theorem 7 (First Difference).

```
⊢ ∀f z. z IN ROC f ∧ is_causal f ⇒
        z_transform (λn. f n − f (n − 1)) z =
        (Cx (&1) − z cpow Cx(−&1)) * z_transform f z
```

4.3 Scaling in the z-Domain or Complex Translation

The scaling property of the z-Transform is useful to analyze communication systems, such as the response analysis of modulated signals in z-domain. If $\mathcal{Z}(f[n])\ z = F(z)$, then two basic types of scaling can be defined as below:

$$\mathcal{Z}(h^n f[n])\ z = F(\frac{z}{h}) \tag{7}$$

$$\mathcal{Z}(\omega^{-n} f[n])\ z = F(\omega z) \tag{8}$$

If h is a positive real number, then it can be interpreted as shrinking or expanding of the z-domain. If h is a complex number with unity magnitude, i.e., $h = e^{j\omega_0}$, then the scaling corresponds to a rotation in the z-plane by an angle of ω_0. On the other hand, multiplication by ω^{-n} (Equation 8) shrinks the $z-$domain. Indeed, in the communication and signal processing literature, it is interpreted as frequency shift or translation associated with the modulation in the time-domain.

We verify the above theorems in HOL Light as follows:

Theorem 8 (Scaling in z-Domain).

```
⊢ ∀f z h. inv h * z IN ROC f ∧ z IN ROC  f ⇒
          z_transform (λn. h cpow Cx (&n) * f n) z =
          z_transform (λn. f n) (inv h * z)
```

Theorem 9 (Scaling in z-Domain (Negative)).

```
⊢ ∀f z w. w * z IN ROC f ∧ z IN ROC f ⇒
          z_transform (λn. w cpow −Cx (&n) * f n) z =
          z_transform (λn. f n) (w * z)
```

4.4 Complex Differentiation

The differentiation property of the z-Transform is frequently used together with shifting properties to find the inverse transform. Mathematically, it can be expressed as:

$$\mathcal{Z}(n * f[n])\ z = -z * (\sum_{n=0}^{\infty} \frac{d}{dz}(f[n]z^{-n})) \tag{9}$$

We prove this property in the following theorem:

Theorem 10 (Complex Differentiation).

```
⊢ ∀f z. &0 < Re z ∧ z IN ROC (Cx (&n) * f n) ⇒
        z_transform  (λn. Cx (&n) * f n) z = −z * infsum (from 0)
           (λn. complex_derivative (λz. f n * z cpow Cx (−&n)) z)
```

The proof of the above theorem requires the properties of complex differentiation, summability and complex arithmetic reasoning.

4.5 Complex Conjugation

The complex conjugation property provides the ease to manipulate the z-Transform of conjugated functions. The mathematical form of this property is as follows:

$$\mathcal{Z}(f^*[n])\ z = F^*(z^*) \tag{10}$$

where $f^*[n]$ represents the complex conjugate of function $f[n]$. The corresponding formal form of the complex conjugation is given as follows:

Theorem 11 (Complex Conjugation).

```
⊢ ∀f z. cnj z IN ROC f ⇒
        z_transform (λn. cnj (f n)) z = cnj(z_transform f (cnj z))
```

4.6 The z-Transform of Commonly Used Functions

In real-world applications, the system is usually subject to a set of known input functions depending upon the dynamics and overall output response. It is quite handy to verify the z-Transform of such functions to simplify the reasoning while tackling practical applications using our formalization. In this regard, we verify the z-Transform expressions for most commonly used functions in signal processing and control systems. Table 1 summarizes these functions along with their mathematical form and corresponding z-Transform. In the following, we provide the formal definition and verification of the z-Transform of the Dirac-Delta function only whereas the verification of other functions can be found in the proof script [23].

Function Name	Mathematical Notation	Z-Transform
Dirac-Delta Function	$\delta[n-m]$	z^{-m}
Exponential	$\exp[-\alpha * n]$	$\frac{1}{1-\exp[-\alpha]z^{-1}}$
Complex Constant	a^n	$\frac{1}{1-az^{-1}}$
Sine	$\sin[\omega_0 n]$	$\frac{z^{-1}\sin[\omega_0]}{1-2z^{-1}\cos[\omega_0]+z^{-2}}$
Cosine	$\cos[\omega_0 n]$	$\frac{1-z^{-1}\cos[\omega_0]}{1-2z^{-1}\cos[\omega_0]+z^{-2}}$
Scaled Sine	$a^n \sin[\omega_0 n]$	$\frac{az^{-1}\sin[\omega_0]}{1-2az^{-1}\cos[\omega_0+a^2z^{-2}]}$
Scaled Cosine	$a^n \cos[\omega_0 n]$	$\frac{1-az^{-1}\cos[\omega_0]}{1-a2z^{-1}\cos[\omega_0+a^2z^{-2}]}$

Table 1: z-Transform of Commonly used Functions

Definition 14 (Dirac-Delta Function).

```
⊢def delta m = (λn. if n = m then Cx (&1) else Cx (&0))
```

Theorem 12 (The z-Transform of Dirac-Delta Function).

```
⊢ ∀z n. z_transform (delta m) z = inv z pow m
```

5 Formalization of Difference Equations

A difference equation characterizes the behavior of a particular phenomena over a period of time. Such equations are widely used to mathematically model complex dynamics of discrete-time systems. Indeed, a difference equation provides a formula to compute the output at a given time, using present and future inputs and past output as given in the following example:

$$y[k] - 5y[k-1] + 6y[k-2] = 3x[k-1] + 5x[k-2] \tag{11}$$

In the perspective of engineering systems, a difference equation is concerned with the generation of a sequence of control outputs $x[n]$ given a sampled sequence of the system inputs $y[n]$. Generally, it is important to determine the control output at a sample instance n based on the sampled system input at the sample instance n and a finite number of previous sampled outputs. Mathematically, it can be written as follows:

$$y[n] = f(x[n], x[n-1], x[n-2], \ldots, x[n-m], y[n-1], y[n-2], \ldots, y[n-k]) \tag{12}$$

There is an infinite number of ways the $m + k - 1$ values on the right-hand side of the above equation can be combined to form $y(n)$. We consider the practical case where the right-hand side of the above equation involves a linear combination of the past samples of the outputs and control inputs, which can be described as follows:

$$y[n] = \sum_{i=1}^{N} \alpha_i y[n-i] + \sum_{i=0}^{M} \beta_i x[n-i] \tag{13}$$

where α_i and β_i are input and output coefficients. The output $y[n]$ is a linear combination of the previous N output samples, the present input $x[n]$ and M previous input samples. Here, α_i and β_i are considered as constants (either complex ($\mathbb{C}$) or real ($\mathbb{R}$)) due to which the Equation (13) is called Linear Constant Coefficient Difference Equation (LCCDE). For a given N^{th} order difference in terms of a function $f[n]$, its z-Transform is given as follows:

$$\mathcal{Z}(\sum_{i=0}^{N} \alpha_i f[n-i])\ z = F(z) \sum_{i=0}^{N} \alpha_i z^{-i} \tag{14}$$

Applying the z-Transform on both sides of Equation (13) results in an important mathematical form describing the relation among the coefficients of $x[n]$ and $y[n]$, called *transfer function* or *system function*, given as follows:

$$H(z) = \frac{Y(z)}{X(z)} = \frac{\sum_{i=0}^{M} \beta_i z^{-i}}{1 - \sum_{i=1}^{N} \alpha_i z^{-i}} \tag{15}$$

In order to build the reasoning support for LCCDE in HOL Light, we formalize the Nth difference as follows:

Definition 15 (N^{th} Difference).

```
⊢def ∀N alst f x. nth_difference alst f N x =
                    vsum (0..N) (λt. EL t alst * f (x - t))
```

The function `nth_difference` accepts the order (N) of the difference equation, a list of coefficients `alst`, function `f` and the variable x. It utilizes the functions `vsum s f` and `EL i L`, which return the vector summation and the ith element of a list L, respectively, to generate the difference equation corresponding to the given parameters.

Next, we formalize a general LCCDE (i.e., Equation (13)) as follows:

Definition 16 (Linear Constant Coefficient Difference Equation (LCCDE)).

```
⊢def ∀y M x N n. LCCDE x y alist blist M N n ⇔
                  y n = nth_difference alist y M n +
                        nth_difference blist x N n
```

Now equipped with these formal definitions, our next step is to verify the z-Transform of the Nth-difference (Definition 15) which is one of the most important results of our formalization.

Theorem 13 (z-Transform of Nth-Difference).

```
⊢ ∀f lst N z. z IN ROC f ∧ is_causal f ⇒
               z_transform (λx. nth_difference lst f N x) z =
               z_transform f z * vsum (0..N)
                                  (λn. z cpow -Cx (&n) * EL n lst)
```

The proof of Theorem 13 is based on induction on the order of the difference and Theorems 2 and 4 along with the following important lemma about the summability of Nth-difference equation:

Lemma 4 (Summability of Difference Equation).

```
⊢ ∀N a_lst f. z IN ROC f ∧ is_causal f ⇒
               z IN  ROC (λx. nth_difference a_lst f N x)
```

In order to verify the transfer function of the LCCDE (Equation (15)), we need to ensure that the input and output functions should be causal as described in Section 3. Another important requirement is to ensure that there are no values of z for which the denominator is 0, such values are called poles of that transfer function. We package these conditions in the following definitions:

Definition 17 (Causal System Parameters).

```
⊢def is_causal_lccde x y ⇔  is_causal x ∧ is_causal y
```

Definition 18 (LCCDE ROC).

```
⊢def ∀x y M alst. LCCDE_ROC x y M alst =
                  (ROC x) INTER (ROC y) DIFF
                  {z | Cx (&1) - vsum (1..M)
                      (λn. EL n alst * z cpow -Cx (&n)) =  Cx(&0)}
```

Here, the function `is_causal_lccde` takes two parameters, i.e., input and output, and ensures that both of them are causal. In Definition 18, `LCCDE_ROC` specifies the region of convergence of the input and output functions, which is indeed the intersection of `ROC x` and `ROC y`, excluding all poles of the transfer function. The function `DIFF` represents the difference of two sets, i.e., $A \setminus B = \{z \mid z \in A \wedge z \notin B\}$.

Next, we present the formal verification of the transfer function as given in Equation 15.

Theorem 14 (LCCDE Transfer Function).

```
⊢ ∀x y alst blst M N.
        z IN LCCDE_ROC x y M alst ∧
        is_causal_lccde x y ∧
        (∀n. LCCDE x y alist blist M N n) ⇒
        z_transform y z / z_transform x z =
        vsum (0..N) (λn. z cpow -Cx (&n) * EL n blst) /
        (Cx (&1) - vsum (1..M) (λn. z cpow -Cx(&n) * EL n alst))
```

The first and second assumptions describe the region of convergence for LCCDE and the causality of the input and output. The last assumption gives the time-domain model of the LCCDE. The proof of this theorem is mainly based on the properties of the z-Transform such as linearity (Theorem 4), time-delay (Theorem 6) and summability of difference equation (Lemma 4). This is a very useful result to simplify the reasoning for the LCCDE of any order.

6 Uniqueness of the z-Transform

One of the most critical aspects of transformation based analysis of discrete-time systems is to be able to obtain the time-domain expressions from z-domain parameters. The inverse transformation is very important to reliably deduce the properties of the underlying system because the actual implementation is done in the time-domain. The inversion of bilateral z-Transform $X(z)$ to its corresponding time domain function $x[n]$ is not unique due to the existence of infinitely many ROCs for one function. However, the uniqueness of unilateral z-Transform (that we have formalized in our work) can be proved considering the nature of the ROC which is always the exterior region of a circle as shown in Figure 3. Mathematically, the uniqueness of the z-Transform can be described as follows:

$$\mathcal{Z}(f[n]) = \mathcal{Z}(g[n]) \Leftrightarrow f = g \tag{16}$$

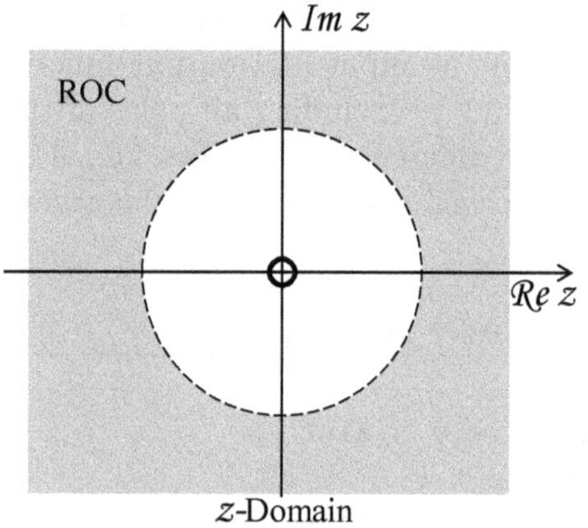

Figure 3: Region of Convergence (ROC) for Inverse z-Transform

The proof of the uniqueness of the z-Transform can be divided into two subgoals, i.e., forward and backward implications as follows:

$$f = g \Longrightarrow \mathcal{Z}(f[n]) = \mathcal{Z}(g[n]) \tag{17}$$

$$\mathcal{Z}(f[n]) = \mathcal{Z}(g[n]) \Longrightarrow f = g \tag{18}$$

The first subgoal is straight forward and can be proved by the definition of the z-Transform. However, the second subgoal requires the reconstruction of the original function $f[n]$ from the transformed function $\mathcal{Z}(f[n])$ or $(F(Z))$. There are two main methods for obtaining such reconstruction: First, a sequence that consists of the coefficients of the Laurent series of $F(z)$, which is given by the following equation [10]:

$$f(k) = \frac{1}{2\pi i} \oint_C F(z)^{k-1} dz \quad (k = 0, 1, 2, ...) \tag{19}$$

where the path of integration C is a circle of radius $r > \rho$ traversed in the anticlockwise direction. The second method involves the higher-order complex derivative of the infinite summation (i.e., the z-Transform $F(z)$) at origin ($z = 0$), given as follows [10]:

$$f(k) = \frac{1}{k!}(\frac{d^k}{dz^k}F(\frac{1}{z}))_{z=0} \quad (k = 0, 1, 2, ...) \tag{20}$$

Interestingly, the multivariate analysis libraries of HOL Light are rich enough to tackle both proofs involving the path integrals and the higher-order complex derivatives of a complex series. However, we have chosen the second methods due to the availability of some important lemmas in HOL Light as described in the sequel.

6.1 Formal Proof of Uniqueness

The unilateral z-Transform is unique for the ROC which forms an exterior of a circle excluding the centre as shown in Figure 3. We formally define the exterior region of a circle as follows:

Definition 19 (Exterior of Circle).

```
⊢def ∀s. exterior_circle s ⇔
        (∃R. &0 < R ∧ (∀z. R < dist (z,Cx (&0)) ⇒  z IN s))
```

where `exterior_circle` accepts a set of complex elements ($\texttt{s} : (\texttt{real}^2 \rightarrow \texttt{bool})$) which forms an exterior region of a circle.

We verify three important properties describing the relation between the ROC of the z-Transform and the exterior of a circle.

- If the ROCs of the two functions f and g are exterior regions of a circle, then the intersection of their ROCs will also form an exterior circle.

```
⊢ ∀f g. exterior_circle (ROC f) ∧ exterior_circle (ROC g) ⇒
        exterior_circle (ROC f INTER ROC g)
```

- If a function f is summable, then its ROC will always form an exterior region of a circle.

```
⊢ ∀f. summable (from 0) f ⇒  exterior_circle (ROC f)
```

- If a function f is decaying over time, then its ROC will always be an exterior region of a circle.

```
⊢ ∀f c N. c < &1 ∧ (∀n. n ≥ N ⇒
          norm (f (SUC n)) ≤ c * norm (f n)) ⇒
          exterior_circle (ROC f)
```

We next prove the inverse transform function given in Equation 20.

Theorem 15 (Inverse z-Transform).

```
⊢ ∀f n. exterior_circle (ROC f) ⇒
        f n = higher_complex_derivative n
              (λz. z_transform f (inv z)) (Cx(&0)) / Cx(&(FACT n))
```

where the proof of Theorem 15 is done using the higher-order derivatives of a power series which is already available in HOL Light, as given in the following form:

Lemma 5 (Higher-Order Derivative of Power Series).

```
⊢ ∀f c r n k. &0 < r ∧ n IN k ∧
              (∀w. dist (w,z) < r ⇒
                   ((λi. c i * (w - z) pow i) sums f w) k) ⇒
              higher_complex_derivative n f z / Cx(&(FACT n)) = c n
```

Finally, we prove the uniqueness of the z-Transform based on Theorem 15 and Lemma 5 along with some complex arithmetic reasoning.

Theorem 16 (Uniqueness of the z-Transform).

```
⊢ ∀f g. exterior_circle (ROC f) ∧ exterior_circle (ROC g) ⇒
        (z_transform f = z_transform g ⇔  f = g)
```

6.2 Initial Value Theorem of the z-Transform

In many situations, it is desirable to compute the initial value of the function from its z-Transform. This is mainly achieved by using the famous initial value theorem of the z-Transform, which states that if the z-Transform of $x[k]$ is $X(z)$ and if $\lim_{z\to\infty} X(z)$ exists, then the initial value of $x[k]$ (i.e., $x[0]$) can be obtained from the following limit:

$$x(0) = \lim_{z\to\infty} X(z) \tag{21}$$

Theorem 17 (Initial Value Theorem).

```
⊢ ∀f. exterior_circle (ROC f) ⇒
      f 0 = lim at_infinity (λz. z_transform f z)
```

The proof of Theorem 17 is mainly based on the concepts about the differentiability, continuity and theory of holomorphic functions, as described in the following two lemmas (which are available in the HOL Light multivariate theory).

Lemma 6 (Complex Differentiability Implies Continuity).

```
⊢ ∀f x. f complex_differentiable at x ⇒  f continuous at x
```

Lemma 7 (Holomorphic Implies Differentiability).

```
⊢ ∀f s x. f holomorphic_on s ∧ open s ∧ x IN s ⇒
          f complex_differentiable at x
```

7 Applications

In order to illustrate the utilization and effectiveness of the reported formalization, we present the formal analysis of a couple of real-world applications namely power converters and digital filters which are widely used systems in the domain of power electronics and digital signal processing, respectively.

7.1 Formal Analysis of Switched-Capacitor Power Converter

In the last decade, very-large scale integrated (VLSI) systems industry has revolutionized many fields of physical sciences and engineering including communication, mobile devices and health-care. However, increased density of integrated chips resulted in high power dissipation which is known as energy crisis in VLSI industry. In order to overcome this issue, power management techniques can be applied at the system, circuit or device level depending on the system complexity and nature of the device operation. The system level power management techniques are used to identify optimal operating conditions by power sensing and power management. DC-DC converter [17] is one of the most important circuit level power management modules which convert an unregulated input DC voltage into an output voltage. Mainly, integrated DC-DC converters can be divided into three classes namely linear regulators, switch mode power converters and switched-capacitor power converters [17]. In this paper, we aim at formal modeling and analysis of switched-capacitor (SC) DC-DC converters due to their robustness and wide application domain [14].

7.1.1 Mathematical Modeling of SC Power Converter

In the design and modeling of any kind of power converter, it is very critical to obtain the transfer function (the input-output relation) to analyze the overall system design, system stability and desired power-gain. Generally, power electronics engineers obtain the power stage transfer function of switch mode and SC power converters using the z-Transform. Figure 4 outlines the system architecture of the interleaved SC power converter. The power stage is a cross-coupled voltage doubler

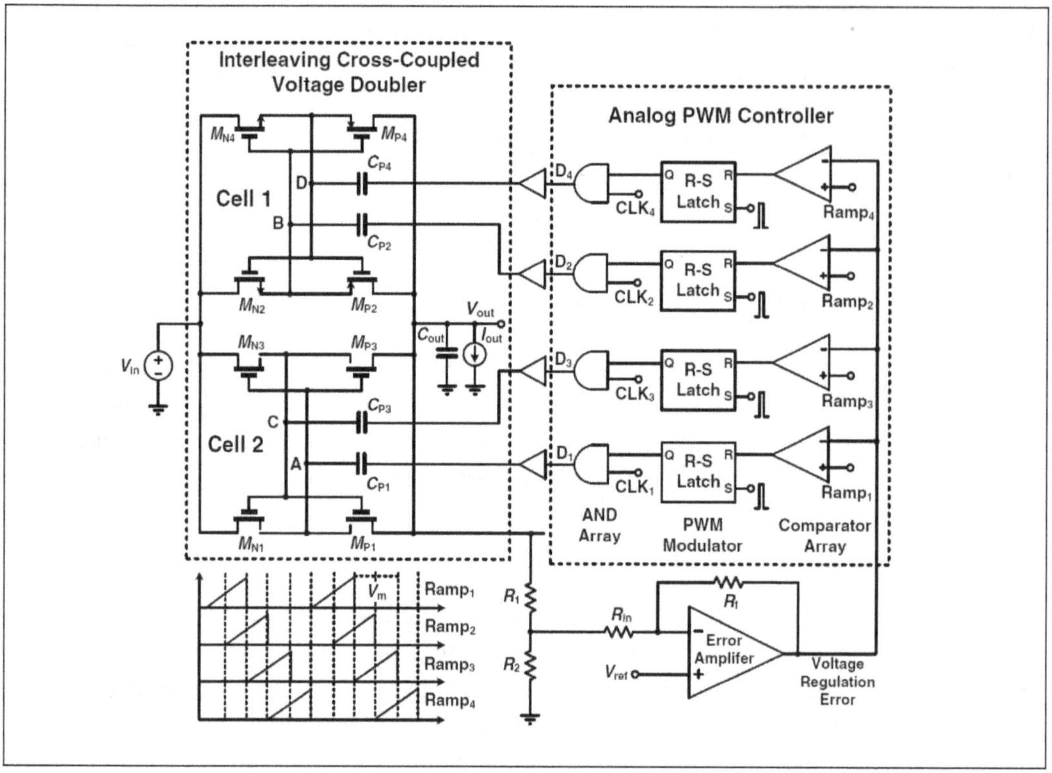

Figure 4: System Architecture of the Interleaved SC Power Converter [17]

that is regulated using an analog pulse-width modulation (PWM) controller. We can briefly describe its operation as follows: Initially, V_{out} is scaled down with the aid of a resistive voltage divider. This scaled voltage is then compared with the desired reference voltage V_{ref} and the corresponding voltage regulation error is determined and amplified by the error amplifier. The output of the error amplifier is then used to determine the output-input ratio of each charge pump sub-cell [17].

In order to derive the transfer function of the cross-coupled voltage doubler, Figure 5 describes the charge and discharge process of one charge pump cell. The charge pump operates in a full charge mode in which the current delivered by the pumping capacitors C_{Pi} at the end of each switching interval drops to a very low level, in comparison to its peak value. Since the two cross-coupled cells do not exchange charge or power at any instant during their operation, they can be modeled as separate elements. Finally, the overall operation can be modelled by the following six equations:

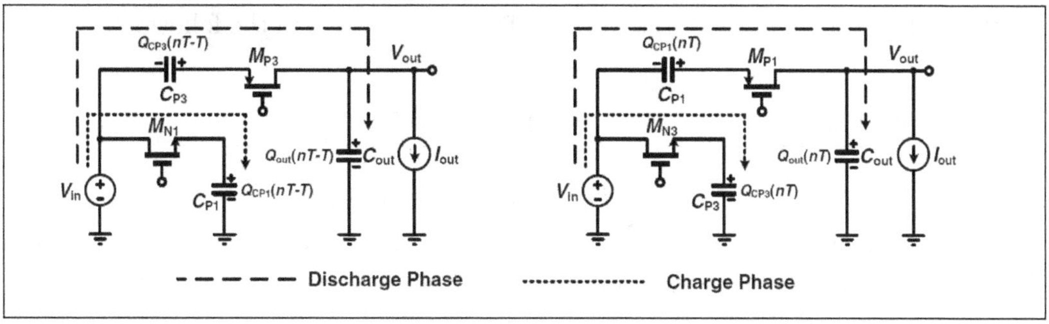

Figure 5: Charge and Discharge Phases for Interleaving SC Power Converter [17]

$$Q_1(n) = C_p(V_{out}(n) - V_{in}(n)) \tag{22}$$

$$Q_3(n) = C_pV_{in}(n) \tag{23}$$

$$Q_{out}(n) = C_{out}V_{out}(n) \tag{24}$$

$$Q_1(n-1) = C_pV_{in}(n-1) \tag{25}$$

$$Q_3(n-1) = C_p(V_{out}(n-1) - V_{in}(n-1)) \tag{26}$$

$$Q_{out}(n-1) = C_{out}V_{out}(n-1) \tag{27}$$

where Q_i, represents the charge stored at different nodes in the circuit, whereas V_{in} and V_{out} represent the input and output of the voltage doubler. The total charge transfer can be described as follows:

$$2 * (Q_1(n-1) - Q_1(n) + Q_3(n) - Q_3(n-1)) + Q_{out}(n-1) - Q_{out}(n) = \frac{T_s}{2}[\frac{V_{out}(n-1)}{R_{out}} + \frac{V_{out}(n)}{R_{out}}] \tag{28}$$

where R_{out} is output load resistor.

Using Equations ((22)-(28)) and the z-Transform results in the following transfer function:

$$\frac{V_{out}(z)}{V_{in}(z)} = \frac{4C_p(1+z^{-1})}{(2C_p - C_{out} + \frac{Ts}{2R_{out}})(\frac{(2C_p+C_{out}-\frac{Ts}{2R_{out}})}{(2C_p-C_{out}+\frac{Ts}{2R_{out}})} + z^{-1})} \tag{29}$$

Finally, letting $z = 1$ and $Ts = 0$, results in the DC conversion gain which should be consistent with the gain of an ideal voltage doubler, i.e., 2, as follows:

$$\left[\frac{V_{out}(z)}{V_{in}(z)}\right]_{z=1,Ts=0} = 2 \tag{30}$$

7.1.2 Formal Verification of the Transfer Function and DC Conversion Gain

Our main goal is to verify the transfer function of the voltage doubler (Equation (29) and the DC conversion gain (Equation (30), which are two critical requirements in the correct operation of the interleaved SC power converters. We formalize Equations ((22)-(28)) in HOL Light as follows:

Definition 20 (Voltage Doubler Model).

```
⊢def sc_voltage_doubler Q1 Q3 Qout Cp Cout Vin Vout ⇔
     (∀n. Q1 n = Cp * (Vout n − Vin n) ∧
     Q3 n = Cp * Vin n ∧ Qout n = Cout * Vout n ∧
     Q1 (n − 1) = Cp * Vin (n − 1) ∧
     Q3 (n − 1) = Cp * (Vout (n − 1) − Vin (n − 1)) ∧
     Qout (n − 1) = Cout * Vout (n − 1))
```

where the three variables `Q1`, `Q3` and `Qout` represent the values of the stored charge at different nodes. The parameters `Cp` and `Cout` represent the capacitors, whereas `Vin` and `Vout` represent the input and output voltages, respectively. The function `sc_voltage_doubler` returns the corresponding model of the voltage doubler corresponding to Equations ((22)-(27)). We next formally define the total transfer charge (Equation (28) as follows:

Definition 21 (Total Transfer Charge).

```
⊢def ∀Q1 Q3 Qout Ts Vout n Rout.
      total_charge_transfer Q1 Q3 Qout Vout Rout Ts n ⇔
      Cx(&2) * (Q1 (n − 1) − Q1 n + Q3 n − Q3 (n − 1)) +
      Qout (n − 1) − Qout n =
      Cx Ts / Cx(&2) * (Vout (n − 1) − Vout n) / Cx Rout
```

We next verify the transfer function of the SC Voltage doubler as follows:

Theorem 18 (SC Voltage Doubler Transfer Function).

```
⊢ ∀Q1 Q3 Qout n Vin Vout Cp Cout Ts Rout z.
    [A1] sc_voltage_doubler Q1 Q3 Qout Cp Cout Vin Vout ∧
    [A2] total_charge_transfer Q1 Q3 Qout Vout Rout Ts n ∧
    [A3] sc_parameters_constraints Cp Rout Cout Ts z ∧
    [A4] z IN ROC Vin ∧ z IN ROC Vout ∧
    [A5] is_causal Vin ∧ is_causal Vout ⇒
         transfer_function Vin Vout z =
         (Cx(&4) * Cp * (Cx(&1) + z cpow −Cx(&1))) /
         ((Cx(&2)* Cp − Cout + Cx Ts / (Cx(&2) * Cx Rout)) *
         ((Cx(&2)* Cp + Cout − Cx Ts / (Cx(&2) * Cx Rout)) /
         (Cx(&2) * Cp − Cout +
                  Cx Ts / (Cx(&2) * Cx Rout)) + z cpow −Cx(&1)))
```

where assumptions A1 and A2 describe the function of the SC voltage doubler and total transfer charge, respectively. The assumption A3 describes the constraints among the parameters of the SC voltage doubler so that the transfer function is well defined (i.e., there are no poles at which it becomes undefined). The assumptions A4 and A5 ensure that the input and output voltages are causal functions and form valid ROCs. The function `transfer_function` takes an input function `x`, an output function `y` and a z-domain parameter `z:complex` and returns the z-domain transfer function `z_transform y z / z_transform x z`.

Finally, we utilize Theorem 18 to verify the corresponding DC conversion gain of the voltage doubler configuration as follows:

Theorem 19 (SC Voltage Doubler Transfer Function).

```
⊢ ∀Q1 Q3 Qout n Vin Vout Cp Cout Rout.
    [A1] sc_voltage_doubler Q1 Q3 Qout Cp Cout Vin Vout ∧
    [A2] total_charge_transfer Q1 Q3 Qout Vout Rout (&0) n ∧
    [A3] sc_parameters_constraints Cp Rout Cout (&0) z ∧
    [A4] summable (from 0) Vin ∧ summable (form 0) Vout
    [A5] is_causal Vin ∧ is_causal Vout ⇒
         transfer_function Vin Vout Cx(&1) = Cx(&2)
```

In this application, we present the design of an interleaved cross-coupled SC voltage doubler, which is regulated using an analog PWM control scheme. We demonstrate the use of our formalization of the z-Transform and its properties by the formal modeling and verification of the SC interleaved cross-coupled SC voltage doubler. Similar analysis steps can be followed to analyze more converter configurations such as the monolithic SC power converter and the charge pump [17].

7.2 Formal Analysis of Infinite Impulse Response Filters

Digital filters are fundamental components of almost all signal processing and communication systems. The main functionality of such components are to: 1) limit a signal within a given frequency band; 2) decompose a signal into multiple bands; and 3) model the input-output relation of complicated systems such as mobile communication channels and radar signal processing. Digital filters can be used for the performance specifications which are very difficult to achieve by analog filters. Moreover, the functionality of digital filters can be controlled using software applications. Due to these features, such filters are widely used in adaptive filtering applications in telecommunications, speech recognition and biomedical devices.

An impulse response of a system describes its behavior under an external change (mathematically, this describes the system response when the Dirac-Delta function is applied as an input [21]). Infinite impulse response (IIR) filters have an impulse response function which is non-zero over an infinite length of time. In practice, IIR filters are implemented using the feedback mechanism, i.e., the present output depends on the present input and all previous input and output samples. Such an architecture requires delay elements due to the discrete nature of input and output signals. The highest delay used in the input and the output function is called the order of the filter. The time-domain difference equation describing a general M^{th} order IIR filter, with N feed forward stages and M feedback stages, is shown in Figure 6.

Mathematically, it can be described as:

$$y[n] - \frac{1}{\alpha_0} \sum_{i=1}^{M} \alpha_i y[n-i] = \sum_{i=0}^{N} \beta_i x[n-i] \tag{31}$$

where α_i and β_i are input and output coefficients (Note that, $\alpha_0 = 1$ in most practical situations [21]). In case of a time-invariant filter, α_i and β_i are considered constants (either complex ($\mathbb{C}$) or real ($\mathbb{R}$)) to obtain the filter response according to the given specifications.

Our main objective is to formally verify the frequency response of an IIR filter which is given as follows:

$$H(\omega) = \left[\frac{Num}{Den}\right] * \exp\left(j * Arg\left[\frac{Num}{Den}\right]\right) \tag{32}$$

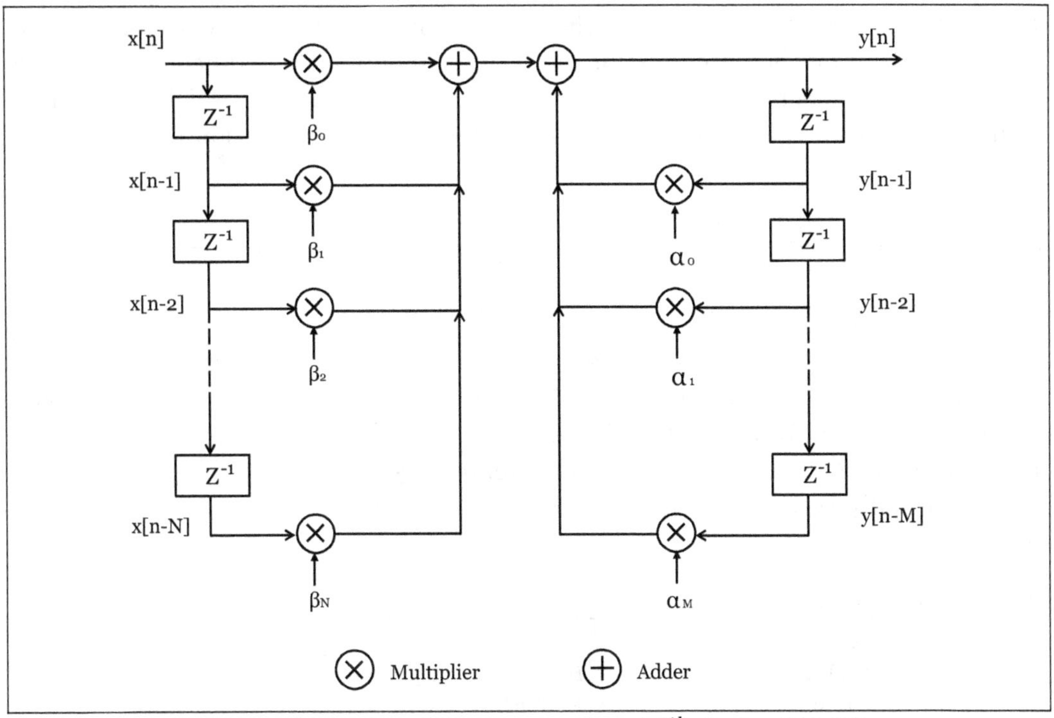

Figure 6: Generalized Structure of an Mth Order IIR Filter

where:

$$Num = \parallel (\sum_{i=0}^{N} \beta_i cos(i\omega)) - j(\sum_{i=0}^{N} \beta_i sin(i\omega)) \parallel \tag{33}$$

$$Den = \parallel (1 - \sum_{i=1}^{M} \alpha_i cos(i\omega)) + j(\sum_{i=1}^{M} \alpha_i sin(i\omega)) \parallel \tag{34}$$

Note that $\parallel . \parallel$ represents complex norm and $H(\omega)$ represents the complex frequency response of the filter. The function $Arg(z)$ represents the argument of a complex number [21]. Equation 32 can be derived from the transfer function $H(z)$ by mapping z on the unit circle, i.e., $z = exp(j * \omega)$. The parameter ω represents the angular frequency.

7.2.1 Formal Verification of the Frequency Response of the IIR Filter

Based on the above description of the IIR filter, our next move is to verify the frequency response (Equation (32)), which mainly involves two major steps, i.e., formal description of the model and the verification of the frequency response which is mainly based on the derivation of the transfer function. The difference equation (Equation (31)) describing the dynamics of IIR is similar to the LCCDE (i.e., Equation (13)). So we can model the IIR filter using the formalization of LCCDE as follows:

Definition 22 (IIR Model).

```
⊢def ∀y M x N n. iir_model x y a_list b_list M N n =
                   LCCDE x y alist blist M N n
```

The function `iir_model` defines the dynamics of the IIR structure in a relational form. It accepts the input and output signals ($\mathtt{x}, \mathtt{y} : \mathbb{N} \rightarrow \mathbb{C}$), a list of input and output coefficients ($\mathtt{a\_lst}, \mathtt{b\_lst} : (\mathbb{C}(\mathtt{list}))$), the number of feed forward and feedback stages (`N`, `M`) and a variable `n`, which represents the discrete time.

We formally verify the frequency response of the filter given in Equation 32 as follows:

Theorem 20 (IIR Frequency Response).

```
⊢ ∀x y N blst M w alst.
    cexp(j * w) IN LCCDE_ROC x y M alst ∧ is_causal_lccde x y ∧
    (∀n. iir_model x y alst blst M N n) ∧
    ¬(z_transform x (cexp (j * w)) = Cx(&0)) ⇒
    (let H = transfer_function x y (cexp (j * w)) and
    num_real = vsum (0..N) (λn. ccos (Cx(&n) * w) * EL n blst) and
    num_im = j * vsum (0..N) (λn. csin (Cx(&n) * w) * EL n blst) and
    denom_real = Cx(&1) - vsum (1..M)
                          (λn. ccos (Cx(&n) * w) * EL n alst) and
    denom_im = j * vsum (1..M) (λn. csin (Cx(&n) * w) * EL n alst) in
    H = Cx(norm (num_real - num_im) / norm (denom_real + denom_im)) *
        cexp(j * Cx(Arg((num_real - num_im) / (denom_real + denom_im)))))
```

where `cexp` and `Arg` represent complex exponential and argument of a complex number, respectively. The verification of the above theorem is mainly based on Theorem 14 and tedious complex analysis involving complex norms and transcendental functions.

This completes our formal analysis of a generalized IIR filter which demonstrates the effectiveness of the proposed theorem proving based approach to reason about practical discrete-time linear systems. The availability of the z-Transform properties greatly simplified the verification of the transfer function and frequency response. Moreover, Theorem 20 provides the generic results due to the universal quantification over the system parameters such as input and output coefficients (α_i and β_k, where

$i = 0, 1, 2, \ldots, M$ and $k = 1, 2, \ldots, N$), which is not possible in case of simulation based analysis of an IIR filter.

Thanks to the rich multivariate libraries of the HOL Light theorem prover, we have been able to formalize the z-Transform, which is an important tool to model discrete-time linear systems. The overall formalization reported in this paper consists of around 2000 lines of the HOL Light script. Indeed the underlying formalization of the z-Transform including its properties and the uniqueness took around 1700 lines of code, wheras the analysis of both applications took around 300 lines of code. The main contribution of formalizing the z-Transform in HOL can be seen as twofold: 1) To demonstrate the effectiveness of current state-of-the-art technology in theorem proving to formalize the fundamentals of engineering mathematics; 2) To build a formal framework which can be used to reason about the analytical properties of discrete-time systems in the time and frequency domain. Mostly the Laplace transform transfer functions are converted into z-domain to evaluate interesting properties and to obtain corresponding time-domain equations. The main reason behind this choice is the difficulty to obtain the inverse Laplace transform and issues about its uniqueness. In this perspective, the formalization of the z-Transform can also be used to analyze the continuous-time systems using the Biliear Transform, which is yet to be formalized in higher-order logic.

Note that the verification of the properties of the z-Transform had to be done in an interactive way due to the undecidable nature of higher-order logic. The main advantages of this long process are the accuracy of the verified results and digging out all the hidden assumptions, which are usually not mentioned in the textbooks and engineering literature. We believe that this is a one-time investment as the verification of applications becomes quite easy due to the availability of already verified properties of the z-Transform. As mentioned in [2], the availability of fundamental libraries of mathematics can attract mathematicians to use interactive theorem proving for verifying key lemmas in their work, so as in the case of engineers.

8 Conclusion and Future Directions

In this paper, we reported the formal analysis of discrete-time systems using the z-Transform which is one of the most widely used transform methods in signal processing and communication engineering. We leveraged upon the high expressiveness and the soundness of the HOL Light theorem prover to formalize the fundamental properties (e.g., time delay, time advance, complex translation and initial value theorem) of the z-Transform and linear constant coefficient difference equations. We also discussed and presented a proof of the uniqueness of the z-Transform which is required to transform z-domain expressions in the time-domain. Finally, in order to demonstrate the effectiveness of the developed formalization, we presented the formal analysis of a switched capacitor voltage doubler and a generalized infinite impulse response filter. Our reported work can be considered as a step towards an ultimate goal of using theorem provers in the design and analysis of systems from different engineering and physical science disciplines (e.g., signal processing, control systems, biology, optical and mechanical engineering).

In future, we plan to use the formalization of the z-Transform to verify the properties of photonic filters [5, 18] and discrete-time fractional order systems [27, 25]. In both these applications, our current formalization can be substantially used in its current state. However, the analysis of fractional order systems require more formalization of discrete fractional derivates based on the theory of special functions (e.g., Gamma Function [26, 13]). Another interesting direction is the development of a formal link between the z-Transform and the signal-flow-graph [24], which is a complementary technique to obtain the transfer functions of various engineering systems [4, 3]. Indeed such a formal link will provide a framework to use our formalization to reason about graphical models of signal processing and control systems often realized in MATLAB Simulink.

References

[1] A.S. Alfa. *Queueing Theory for Telecommunications - Discrete Time Modelling of a Single Node System.* Springer, 2010.

[2] J. Avigad and J. Harrison. Formally Verified Mathematics. *Communications of the ACM*, 57(4):66–75, 2014.

[3] S.M. Beillahi, U. Siddique, and S. Tahar. Formal Analysis of Power Electronic Systems. In *Formal Methods and Software Engineering*, volume 9407 of *Lecture Notes in Computer Science*, pages 270–286. Springer, 2015.

[4] S.M. Beillahi, U. Siddique, and S. Tahar. Formal Analysis of Engineering Systems Based on Signal-Flow-Graph Theory. In *Numerical Software Verification*, volume 10152 of *Lecture Notes in Computer Science*, pages 31–46. Springer, 2017.

[5] L.N. Binh. *Photonic Signal Processing: Techniques and Applications.* Optical Science and Engineering. Taylor & Francis, 2010.

[6] P.S.R. Diniz, E.A.B. Da Silva, and S.L. Netto. *Digital Signal Processing: System Analysis and Design.* Cambridge University Press, 2002.

[7] S. Elaydi. *An Introduction to Difference Equations.* Springer, 2005.

[8] S. Fadali and A. Visioli. *Digital Control Engineering: Analysis and Design.* Academic Press, 2012.

[9] G. Gonthier. Engineering Mathematics: The Odd Order Theorem Proof. In *Proceedings of the ACM SIGPLAN-SIGACT Symposium on Principles of Programming Languages*, pages 1–2. ACM, 2013.

[10] U. Graf. *Applied Laplace Transforms and z-Transforms for Scientists and Engineers: A Computational Approach using a Mathematica Package.* Birkhäuser Basel, 2012.

[11] T. Hales. *Dense Sphere Packings: A Blueprint for Formal Proofs*, volume 400 of *London Mathematical Society Lecture Note Series.* Cambridge University Press, 2012.

[12] J. Harrison. The HOL Light Theory of Euclidean Space. *Journal of Automated Reasoning*, 50(2):173–190, 2013.

[13] J. Harrison. Formal Proofs of Hypergeometric Sums. *Journal of Automated Reasoning*, 55(3):223–243, 2015.

[14] A. Schultz J. Li, C.R. Sullivan. Coupled-Inductor Design Optimization for Fast-Response Low-Voltage DC-DC Converters. In *IEEE Applied Power Electronics Conference and Exposition*, volume 2, pages 817–823, 2002.

[15] E.I. Jury. *Theory and Application of the Z-Transform Method.* Wiley, 1964.

[16] B.P. Lathi. *Linear Systems and Signals.* Oxford University Press, 2005.

[17] D. Ma and R. Bondade. *Reconfigurable Switched-Capacitor Power Converters.* Springer, 2013.

[18] S. Mandal, K. Dasgupta, T.K. Basak, and S.K. Ghosh. A Generalized Approach for Modeling and Analysis of Ring-Resonator Performance as Optical Filter. *Optics Communications*, 264(1):97 – 104, 2006.

[19] Mathematica: Signal Processing Functions. `http://reference.wolfram.com/mathematica/guide/SignalProcessing.html`, 2018.

[20] MathWorks: Signal Processing Toolbox. `http://www.mathworks.com/products/signal/`, 2018.

[21] A.V. Oppenheim, R.W. Schafer, and J.R. Buck. *Discrete-Time Signal Processing.* Prentice Hall, 1999.

[22] A. Rashid and O. Hasan. On the Formalization of Fourier Transform in Higher-order Logic. In *Interactive Theorem Proving*, volume 9807 of *Lecture Notes in Computer Science*, pages 483–490. Springer, 2016.

[23] U. Siddique. Formal Anlysis of Discrete-Time Linear Systems. `http://hvg.ece.concordia.ca/projects/signal-processing/discrete-time.html`, 2018.

[24] U. Siddique, S.M. Beillahi, and S. Tahar. On the Formal Analysis of Photonic Signal Processing Systems. In *Formal Methods for Industrial Critical Systems*, volume 9128 of *Lecture Notes in Computer Science*, pages 162–177, 2015.

[25] U. Siddique and O. Hasan. Formal Analysis of Fractional Order Systems in HOL. In *Formal Methods in Computer-Aided Design*, pages 163–170. IEEE, 2011.

[26] U. Siddique and O. Hasan. On the Formalization of Gamma Function in HOL. *Journal of Automated Reasoning*, 53(4):407–429, 2014.

[27] U. Siddique, O. Hasan, and S. Tahar. Towards the Formalization of Fractional Calculus in Higher-Order Logic. In *Intelligent Computer Mathematics*, volume 9150 of *Lecture*

Notes in Computer Science, pages 316–324. Springer, 2015.

[28] U. Siddique, M.Y. Mahmoud, and S. Tahar. On the Formalization of Z-Transform in HOL. In *Interactive Theorem Proving*, volume 8558 of *Lecture Notes in Computer Science*, pages 483–498. Springer, 2014.

[29] U. Siddique and S. Tahar. On the Formal Analysis of Gaussian Optical Systems in HOL. *Formal Aspects of Computing*, 28(5):881–907, 2016.

[30] U. Siddique and S. Tahar. Formal Verification of Stability and Chaos in Periodic Optical Systems. *Journal of Computer and System Sciences*, 88:271 – 289, 2017.

[31] D. Sundararajan. *A Practical Approach to Signals and Systems.* Wiley, 2009.

[32] S.H. Taqdees and O. Hasan. Formalization of Laplace Transform Using the Multivariable Calculus Theory of HOL-Light. In *Logic for Programming, Artificial Intelligence, and Reasoning*, volume 8312 of *LNCS*, pages 744–758. Springer, 2013.

[33] X.S. Yang. *Mathematical Modeling with Multidisciplinary Applications.* John Wiley & Sons, 2013.

Received 21 September 2015

Does Negative Mass Imply Superluminal Motion? An Investigation in Axiomatic Relativity Theory

J.X. Madarász, G. Székely
Alfréd Rényi Institute of Mathematics,
Hungarian Academy of Sciences, P.O.Box 127, Budapest 1364, Hungary
madarasz.judit@renyi.mta.hu, szekely.gergely@renyi.mta.hu

M. Stannett
University of Sheffield, Department of Computer Science,
Regent Court, 211 Portobello, Sheffield S1 4DP, United Kingdom
m.stannett@sheffield.ac.uk

Abstract

Formalization of physical theories using mathematical logic allows us to discuss the assumptions on which they are based, and the extent to which those assumptions can be weakened. It also allows us to investigate hypothetical claims, and hence identify experimental consequences by which they can be tested. We illustrate the potential for these techniques by reviewing the remarkable growth in First Order Relativity Theory (FORT) over the past decade, and describe the current state of the art in this field. We take as a running case study the question "*Does negative mass imply superluminal motion?*", and show how a many-sorted first-order theory based on just a few intuitively obvious, but rigorously expressed, axioms allows us to formulate and answer this question in mathematically precise terms.

Key words: axiomatic physics, special relativity, dynamics, tachyon, negative mass, logic

PACS: 03.30.+p, 03.65.Ca

The authors would like to thank the anonymous referees for their helpful and insightful comments.

1 Axiomatization of Physical Theories

Relativity theory has been intrinsically axiomatic since its birth, since Einstein presented his 1905 theory of special relativity as a consequence of two informal postulates [12]. Since then several distinct formal axiomatizations of relativity theories (both special and general) have appeared in the literature (see, e.g., [6] and references therein). More recently, a number of researchers have started working on comparing and connecting these different axiomatizations, as well as developing and improving tools to make this possible [5, 7, 8, 28, 31]. In this paper, we work in the framework developed by the research team/school of Hajnal Andréka and István Németi [6, 1, 4], and illustrate the techniques involved by formulating and investigating the question "*does negative mass imply superluminal motion?*" within that framework.

We have chosen this question for our case study because it illustrates a particularly powerful application of the logical approach, viz. the ability to formulate and reason about concepts about which we do not yet have any experimental experience. In such circumstances the ability to write down formal definitions and make logical deductions is essential. If we can show that a concept leads inexorably to logical paradox, we thereby provide firm evidence that the concept is unphysical. Alternatively, we may discover physically feasible preconditions under which the concept is logically entailed, and this in turn gives the potential to devise relevant experimental tests. For example, we know that simple inelastic collisions between positive-mass slower-than-light particles cannot result in particles moving faster-than-light (tachyons). So any experiment in which tachyons are generated through a simple inelastic collision of slower-than-light particles must entail the existence of negative-mass particles. Conversely, as we show formally below, the possibility of simple inelastic collisions between negative-mass particles necessarily entails the existence of tachyons. As a result, those who wish to refute the possibility of negative mass need only refute the existence of tachyons — and the logical method can again be used to investigate this issue. For example, a common argument against tachyons (and hence against negative-mass particles) is that they would lead to causality violations, but the logical methods espoused in this paper can be used to show that this argument is itself logically flawed — tachyons can exist in relativity theories *without* introducing causality violations [2].

Even though negative mass has never been observed experimentally, physicists have speculated [13] about its existence since at least the 19th century and a considerable amount has been published on the subject. Even in the absence of physical evidence, there are situations where negative mass can be invoked as a useful simplifying concept. For example, negative mass can be used to simplify the dynamics of objects embedded in fluids [11], and similar classical situations where negative mass is a practical concept are discussed by Meyer [24] and Ziauddin [33]. However, there is confusion in the wider literature, because different authors deduce their findings from different background assumptions—this makes it unclear which results can sensibly be combined without accidentally generating logical inconsistencies.

One may contrast the possibility of negative mass to that of negative length. From an intuitive standpoint the length of an object 'ought' to be positive, but when lengths are used in computations it is convenient to use negative and positive values to take account of *orientation*. In contrast, the concept of negative mass is not just a convenient sign notation. It has real empirical, and hence physical, meaning. Something has negative inertial mass if it has negative resistance to changing its state of motion. So if we attempt to slow down a negative-mass body by pushing against it, its velocity will actually increase instead.

In Newtonian theory, mass refers to three distinct concepts. The inertial mass (m_i) of a particle determines how its acceleration is related to the forces acting upon it, its active mass (m_a) gives rise to gravitational fields, and its passive mass (m_p) determines how it is acted upon by gravity. Applying Newton's Third Law to gravitational forces tells us that m_a/m_p is the same for all particles, while the weak equivalence principle (that gravity and acceleration have identical effects) implies that m_i/m_p is a positive constant for any given particle. Choosing units such that $m_a/m_p = m_i/m_p = 1$ therefore allows us to declare that $m_a = m_p = m_i$, and since gravity is observed to be universally attractive one typically assumes that mass is positive.

Hohmann and Wohlfarth [17] note, however, that the experimental basis for these equalities applies only to observable matter, and discuss the possibility that negative mass particles might contribute, at least in part, to the 'dark matter' component of the Universe. For their purposes a particle has negative mass if $m_i/m_p = -1$, but since this violates the weak equivalence principle in relativity theory (which requires $m_i = m_p$) they use a modified version of Einstein gravity in which the geodesics followed by positive masses are defined by one space-time metric and those of negative masses by another, and assume that there is no non-gravitational coupling between the two types of particle (since we could otherwise have observed negative masses already). They note that 'bimetric' models of this kind, which generate asymmetric forces between positive and negative mass particles, are themselves considered by some to be inconsistent [26], and deduce a further constraint on their construction, viz. it is not possible to have gravitational forces of exactly equal strength and opposite direction acting on the two classes of test particle. However, even this result depends on background assumptions, and anti-gravity models are known to exist in which their theorem does not apply [18, 19].

In perhaps the best-known relativistic analysis of negative-mass particles, Bondi [10] successfully constructed a "world-wide nonsingular solution of Einstein's equations containing two oppositely accelerated pairs of bodies, each pair consisting of two bodies of opposite sign of mass". More recently, Belletête and Paranjape [9] have demonstrated in a general relativistic setting that Schwarzchild solutions exist representing matter distributions which are "perfectly physical", despite describing a negative mass geometry outside the matter distribution. Jammer [20] has discussed the historical and philosophical context of negative mass at length. While stressing the fact that no negative-mass particle has yet been observed experimentally, he notes that "no known physical law precludes the existence of negative masses". On the other hand, several unusual (and potentially unphysical) properties of negative mass bodies have been proven using various background theories ranging from Newtonian physics to string theory [27, 16].

However, all of this knowledge is based on assumptions and frameworks which differ from one author to the next, and it is consequently difficult to determine to what extent the various claims are consistent with one another or even exactly what basic assumptions are used in each framework. In the absence of experimental evidence, using a framework where the basic concepts and assumptions are crystal clear is essential, since any inadvertent combination of inconsistent results from the literature would allow us to confirm any claim, no matter how fanciful.

Here we introduce just such a framework to investigate of the consequences of having negative mass bodies. Our framework is delicate enough to formulate precise axioms with clear meanings and formally prove the connection between the existence of negative mass bodies and superluminal ones. At the same time, it is also simple enough to be grasped by a college physics student with only a basic understanding of mathematical logic.

Our results imply the existence of yet another constraint on the existence of negative mass particles. We show formally that if such particles exist, provided they can collide inelastically (i.e. fuse together) with 'normal' particles in collisions that conserve four-momentum, then faster-than-light (FTL) particles must also exist. We prove this by showing how, given any negative mass particle a with known 4-momentum, it is possible to specify a suitable positive mass particle b, such that the inelastic collision of a with b would generate an FTL body. We prove our claims within a general axiomatic logical framework, using axioms that are relevant in both Newtonian and relativistic dynamics. This ensures that we can be certain exactly what is assumed and what is not, and hence confirm the absence of unintended inconsistencies. Moreover, keeping things as general as possible ensures that our results have the widest possible applicability.

Another important feature of our approach is that we explicitly avoid using unstated and potentially unjustifiable assumptions in deriving our results. Avoiding such assumptions, and in particular the blanket assumption that negative-mass particles cannot exist, is important in this context, since it allows us to provide potentially educational explanations as to *why* such phenomena may or may not be physically feasible. In contrast, if we simply assert a priori that negative mass is unphysical, the only answer we can give to the question "why?", is "because we say so". For example, it might be argued informally that the entailed existence of FTL particles, proven in this paper, would itself entail the possibility of causality paradoxes, so that the consequences of negative mass particles are not 'reasonable'. But informal arguments of this nature can be flawed: using our formal approach, we and our colleagues have recently shown that spacetime (of any dimension $1+n$) can be populated with particles and observers in such a way that faster-than-light motion is possible, but this does *not* lead to the 'time travel' situations (so beloved of Star Trek fans) that give rise to causality problems [2]. Consequently, the fact that negative-mass particles entail the existence of FTL particles cannot, of itself, be used to argue logically against their existence.

Formal axiomatization also allows us to address consistency issues and what-if scenarios. It is possible to show, for example, that the consistency of relativistic dynamics with interacting particles having negative relativistic masses follows by a straightforward generalization of the model construction used by Madarász and Székely [23] to prove the consistency of relativistic dynamics and interacting FTL particles, see also [30]. The same approach allows us to derive and prove the validity of key relativistic formulae. For example, we can also show logically that all inertial observers of any particle must agree on the value of $m\sqrt{|1-v^2|}$, where m is the particle's relativistic mass and v its speed ($c=1$). This formally confirms the widely-held 'popular' belief that the observed relativistic mass and momentum of a positive-mass FTL particle must *decrease* as its relative speed increases [21].

We introduce our results in two stages. In Section 2, we show informally that there are several simple ways to create FTL particles using inelastic collisions between positive and negative relativistic mass particles. Then in Section 3, we reconstruct our informal arguments within an axiomatic framework so as to make explicit all the assumptions needed to prove our central claim, that the existence of particles with negative relativistic mass necessarily entails the existence of FTL particles.

In addition to its pedagogic advantages, actively restating and proving our statements formally has a further advantage over the informal approach. The mechanics of proof construction require us to identify all of the tacit assumptions underpinning our informal arguments, thereby revealing which assumptions are relevant and which are unwarranted or unnecessary. Identifying and avoiding those which are unnecessary is itself beneficial, since including different sets of conflicting, but unnecessary, hypotheses could potentially prevent us fusing different areas of physics – e.g., gravity and quantum theory – into a single coherent framework. This is, intriguingly, a task with which automated *interactive theorem provers* [32] are increasingly able to assist, both in terms of proof production and automatic checking of correctness. Indeed, this approach is already leading to the production and machine-verification of non-trivial relativistic theorems [15, 29].

In summary, an obvious didactic benefit of using a formal axiomatic framework for investigating questions such as the one investigated here is the elimination of tacit assumptions. In an axiomatic framework it is clear what is assumed and what is not, as well as where these assumptions are used. (For a more delicate discussion on the epistemological significance of the axiomatic framework used in this paper, see Friend's independent study [14] of this approach.)

2 Generating FTL particles from negative mass particles

Let us assume that particles do indeed exist with negative relativistic mass, and that it is possible for such particles to collide inelastically with 'normal' particles. As we now illustrate informally, the existence of FTL particles (tachyons) follows almost immediately, provided we assume that four-momentum is conserved in such collisions. For simplicity, we take $c = 1$. Throughout this paper, we will always understand 'mass' to mean 'relativistic mass'.

Recall first that the *four momentum* of a particle b is the four-dimensional vector $(m, \mathbf{p})$, where m is its relativistic mass and $\mathbf{p}$ its linear momentum (as measured by some inertial observer whose identity need not concern us, because switching to another observer may change the values of certain quantities but not the main phenomena). Notice also that the particle b is a tachyon if and only if $|m| < |\mathbf{p}|$ (i.e. its observed speed is greater than $c = 1$), and that all inertial observers agree as to this judgement (if one inertial observer considers b to be travelling faster than light, they all do — this is because all inertial observers consider each other to be travelling slower than light relative to one another. For a machine-verified proof of this assertion using our approach, see the work of Stannett and Németi [29]).

In this paper, we concentrate on three special types of collisions so as to emphasize how few background assumptions (e.g., about what kinds of positive mass particles exist) are needed to create FTL particles by inelastically colliding particles of positive and negative mass. We will assume the existence of two colliding particles a and b, where a has negative mass $m < 0$ and b has positive mass $M > 0$, which move along the same spatial line (though possibly in opposite directions). Taking the common line of travel to be the x-axis, positive in the direction of b's travel, the four-momenta of a and b can be written $(m, p, 0, 0)$ and $(M, P, 0, 0)$, respectively, for suitable values of p and P. Assuming that four-momentum is conserved during the collision, the four-momentum of the particle c generated by the fusion of a and b will be $(M + m, P + p, 0, 0)$, and this particle will be a tachyon provided

$$|M + m| < |P + p| \tag{1}$$

If this tachyon has negative mass and positive momentum, it moves in the negative x-direction (it is an unusual property of negative-mass particles that their velocity and momentum vectors point in opposite directions); if it has positive-mass and positive momentum it moves in the positive x-direction. By definition, $M > 0 > m$, and b has both positive mass ($M > 0$) and positive momentum ($P > 0$), since its motion defines the positive x-direction.

2.1 First thought experiment

Suppose a travels slower than light, while b moves at light-speed, so that the four-momenta of a and b can be written $(m, p, 0, 0)$ and $(M, M, 0, 0)$, respectively. According to (1), the particle created by their collision will be a tachyon provided

$$|M + m| < |M + p| \tag{2}$$

There are various ways in which this can happen, depending on the values of m and p (see Fig. 1 and Proposition 1). Notice that $|p| < |m|$ since a travels slower than light.

The case when $|m| = M$, i.e. $M = -m$, is ambiguous irrespective of the velocities of the colliding particles a and b. Since $M + m = 0$ and $|p| < |m| = M$, the linear momentum $M + p$ of the resulting particle c must be positive, even though it has zero relativistic mass. In terms of the space-time diagram (Fig. 2), this means that the particle's worldline is horizontal, i.e. it 'moves' with infinite speed. In these circumstances, the question whether c moves in the positive or negative x-direction is meaningless. However, like other observer-dependent concepts such as simultaneity or the temporal ordering of events, this indeterminacy does not lead to a logical contradiction [23].

2.2 Second thought experiment

Suppose b is stationary, i.e. $P = 0$. By arguments similar to those above, this will result in an FTL particle c whenever $|m| + |\mathbf{p}| > M > |m| - |\mathbf{p}|$, and its direction of travel will be determinate provided $M \neq -m$. See Fig. 3 and Proposition 2.

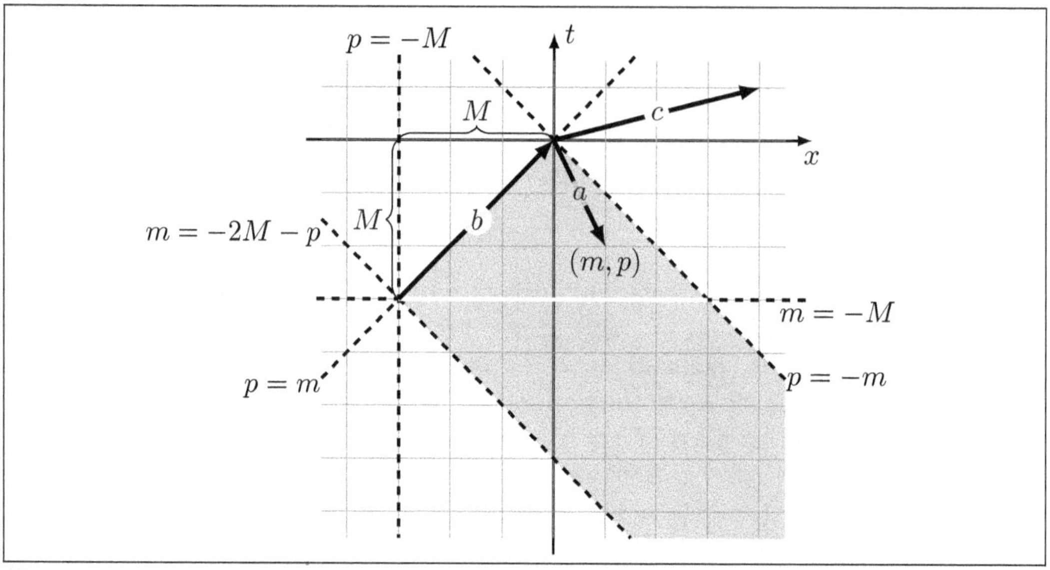

Figure 1: Illustration for generating an FTL particle by colliding a negative relativistic mass particle with a particle moving with the speed of light.

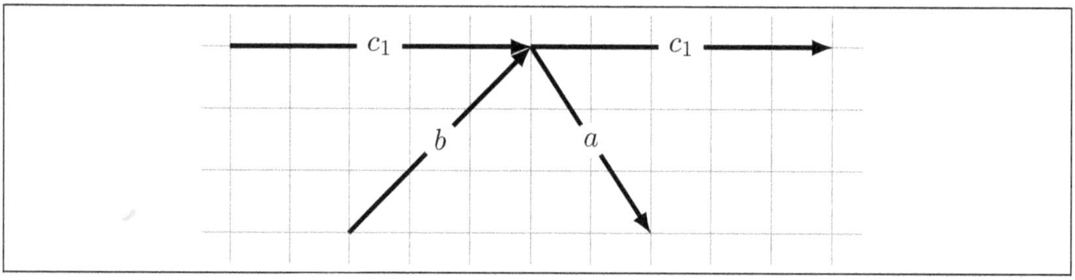

Figure 2: The "inelastic" collision of two particles having opposite relativistic masses is ambiguous in the sense that in this case we have two possible outcomes satisfying the conservation of four-momentum.

2.3 Third thought experiment

Suppose a and b have similar, but oppositely-signed, masses, and that they collide 'head-on' while travelling with equal speeds in opposite directions (relative to some observer, whose identity need not concern us). If the difference in the absolute values of their masses is small relative to their common speed, the resulting particle will be FTL because it will have a small mass relative to its large momentum (which is greater than those of the colliding particles as they have opposite masses); see Proposition 3 for more details.

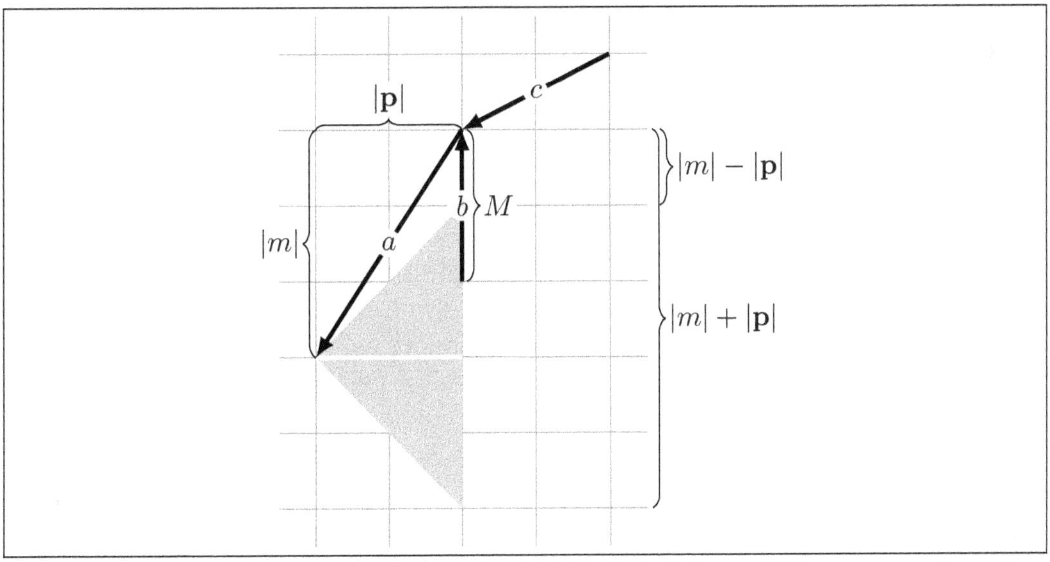

Figure 3: Illustration for generating an FTL particle by colliding a negative relativistic mass particle with a stationary particle of positive relativistic mass.

2.4 FTL particle creations requiring negative relativistic mass

We have seen above that the existence of negative-mass particles implies the existence of FTL particles. Conversely, it is easy to see that an inelastic collision between two slower-than-light particles having positive relativistic masses always leads to a slower-than-light particle. Consequently, the only way in which an inelastic collision between slower-than-light particles can create an FTL particle is if incoming particles can have negative relativistic masses.

In particular, if we impose the condition that such collisions are the only mechanism by which FTL particles can be created, then the existence of FTL particles implies the existence of negative-mass particles. While this suggests that tachyons and negative-mass particles are equally 'exotic', this is, of course, not the case, since the argument that FTL particles require the existence of negative-mass particles relies on the assumption that inelastic collisions are the only mechanism by which FTL particles can be created.

This is by no means a trivial assumption; indeed we have demonstrated elsewhere a consistent model of spacetime in which FTL particles exist, but in which no collisions are posited [2].

3 Axiomatic reconstruction

We have seen three thought experiments in which FTL particles are generated by colliding an arbitrary negative mass particle with an appropriate positive mass one. However, we have not explicitly identified the background assumptions needed to prove our claims concerning these thought experiments. In this section, we dig deeper by identifying these background assumptions; these will turn out to be so general that they are consistent with both relativistic and classical dynamics. To do so, we now reconstruct the above arguments in a precise axiomatic framework, in which each of the used background assumptions will be stated as an explicit axiom. Indeed, making all tacit assumptions explicit can be seen as one of the main advantages of the axiomatic method. Readers interested in the wider context are referred to [23, 29, 21].

3.1 Quantities and Vector Spaces

To formulate the intuitive image above, we need some structure of numbers describing physical quantities such as coordinates, relativistic masses and momenta. Traditional accounts of relativistic dynamics take for granted that the basic number system to be used for expressing measurements (lengths, masses, speeds, etc.) is the field $\mathbb{R}$ of real numbers, but this assumption is far more restrictive than necessary.[1] Instead, we will only assume that the number system is a linearly ordered field Q equipped with the usual constants, zero (0) and one (1); the usual field operations, addition (+), multiplication (·) and their inverses; and the usual ordering (≤) and its inverse; we also assume that the field is *Euclidean*, i.e. positive quantities have square roots. Formally, this is declared as an axiom:

AxEField The structure $\langle Q, 0, 1, +, \cdot, \leq \rangle$ of quantities is a linearly ordered field (in the algebraic sense) in which all non-negative numbers have square roots, i.e. $(\forall x \in Q)((0 \leq x) \Rightarrow (\exists y \in Q)(x = y^2))$.

We write $\sqrt{x}$ for this root, which can be assumed without loss of generality to be both unique and non-negative (regarding machine-verified proofs of this and other relevant claims concerning Euclidean fields, see [29]).

[1]The assumption that $\mathbb{R}$ is the correct number system for expressing lengths (say) is experimentally untestable. Given that we only have access to finitely many measurements, and many of these are (necessarily computable) approximations to 'true' (possibly uncomputable) values, it is not experimentally possible to decide if all non-empty bounded sets of lengths have a supremum, as would be the case if the use of $\mathbb{R}$ were physically necessary.

We choose to use Euclidean fields, as this allows us to refer to 'lengths of vectors' and considerably simplifies the proofs. In practice proofs can generally be modified (by referring instead to 'squared length') to work over any arbitrary ordered field, such as the field of rational numbers. However, that also makes them more complicated. For a paper discussing special relativity in this framework, see [22].

3.2 Inertial particles and observers

We denote the set of physical *bodies* (things that can move) by B. This includes the sets $\mathsf{IOb} \subseteq B$ of **inertial observers**, $\mathsf{Ip} \subseteq B$ of **inertial particles**. Given any inertial observer $k \in \mathsf{IOb}$ and inertial particle $b \in \mathsf{Ip}$, we write $w\ell_k(b) \subseteq Q^4$ for the **worldline** of particle b as observed by k. The coordinates of $\bar{x} \in Q^n$ are denoted by $x_1, x_2, \ldots, x_n$.

The following axiom asserts that the motion of inertial particles are uniform and rectilinear according to inertial observers.

<u>AxIp</u> For all $k \in \mathsf{IOb}$ and $b \in \mathsf{Ip}$, the worldline $w\ell_k(b)$ is either a line, a half-line or a line segment[2].

Suppose observer $k \in \mathsf{IOb}$ sees particle $b \in \mathsf{Ip}$ at the distinct locations $\bar{x}, \bar{y} \in Q^4$. Then its **velocity** according to k is the associated change in spatial component divided by the change in time component,

$$\mathbf{v}_k(b) := \begin{cases} \frac{\mathsf{space}(\bar{x},\bar{y})}{\mathsf{time}(\bar{x},\bar{y})} & \text{if } \mathsf{time}(\bar{x},\bar{y}) \neq 0 \\ \mathsf{undefined} & \text{otherwise} \end{cases}$$

where $\mathsf{space}(\bar{x},\bar{y}) := (x_2 - y_2, x_3 - y_3, x_4 - y_4)$ and $\mathsf{time}(\bar{x},\bar{y}) := x_1 - y_1$. The length[3] of the velocity vector (if it is defined) is the particle's **speed**,

$$v_k(b) := |\mathbf{v}_k(b)|.$$

By AxIp, these concepts are well-defined because $w\ell_k(b)$ lies in a straight line. So the velocities of the considered particles are constants.

If $\mathbf{v}_k(b)$ is defined, we say that b is observed by k to have **finite speed**, and write $v_k(b) < \infty$. The anomalous case $\mathsf{time}(\bar{x},\bar{y}) = 0$ corresponds to a situation where all

[2]Taking $\bar{x}$ and $\bar{y}$ to be of sort Q^4, and λ to be of sort Q, these concepts are defined formally as follows. A *line* is a set of the form $\{\bar{z} \mid (\exists \bar{x}, \bar{y}, \lambda)(\bar{z} = \lambda\bar{x} + (1-\lambda)\bar{y})\}$. A *half-line* is a set of the form $\{\bar{z} \mid (\exists \bar{x}, \bar{y}, \lambda)((0 \leq \lambda) \& (\bar{z} = \lambda\bar{x} + (1-\lambda)\bar{y}))\}$. A *line segment* is a set of the form $\{\bar{z} | (\exists \bar{x}, \bar{y}, \lambda)((0 \leq \lambda \leq 1) \& (\bar{z} = \lambda\bar{x} + (1-\lambda)\bar{y}))\}$.

[3]The **Euclidean length**, $|\bar{x}|$, of a vector $\bar{x}$ is the non-negative quantity $|\bar{x}| = \sqrt{x_1^2 + \cdots + x_n^2}$.

points in $w\ell_k(b)$ are simultaneous from k's point of view, so that k considers the particle to require no time at all to travel from one spatial location to another.

3.3 Collision axioms

In this subsection, we introduce some very simple axioms concerning the dynamics of collisions, and show that the existence of negative relativistic mass implies the existence of faster-than-light (FTL) inertial particles.

Suppose an inertial observer k sees two inertial bodies travelling at finite speed fuse to form a third one at some point $\bar{x}$. In this case, the worldlines of the two incoming particles terminate at $\bar{x}$, while that of the outgoing particle originates there. Formally, we say that an inertial particle b is **incoming** at $\bar{x}$ (according to k) provided $\bar{x} \in w\ell_k(b)$ and $\bar{x}$ occurs strictly later (according to k) than any other point on $w\ell_k(b)$, i.e. $\bar{y} \in w\ell_k(b) \ \& \ \bar{y} \neq \bar{x} \ \Rightarrow \ y_1 < x_1$. **Outgoing** bodies are defined analogously. An **inelastic collision** between two inertial particles a and b (according to observer k) is then a scenario in which there is a unique additional particle $c \in \mathsf{Ip}$ and a point $\bar{x}$ such that a and b are incoming at $\bar{x}$, c is outgoing at $\bar{x}$. We write $\mathsf{inecoll}_k(ab:c)$ to denote that the distinct inertial particles a and b **collide inelastically**, thereby generating inertial particle c (according to observer k). The **relativistic mass** of inertial particle b according to observer k is denoted by $m_k(b)$.

<u>ConsFourMomentum</u> Four-momentum is conserved in inelastic collisions of inertial particles according to inertial observers, i.e.

$$\begin{aligned} &\mathsf{inecoll}_k(ab:c) \Rightarrow \\ &\quad m_k(c) = m_k(a) + m_k(b) \quad \& \\ &\qquad m_k(c)\mathbf{v}_k(c) = m_k(a)\mathbf{v}_k(a) + m_k(b)\mathbf{v}_k(b) \end{aligned}$$

The next axiom, AxInecoll, states that inertial particles moving with finite speeds can be made to collide inelastically in any frame in which their relativistic masses are not equal-but-opposite. Since a collision of particles having equal but opposite relativistic masses does not lead to an inelastic collision according to our formal definition, we do not include this case in this axiom (this does not mean that such particles cannot collide, just that such a collision will not comply with our definition of inelasticity in the associated frame because the third participating particle has infinite speed).

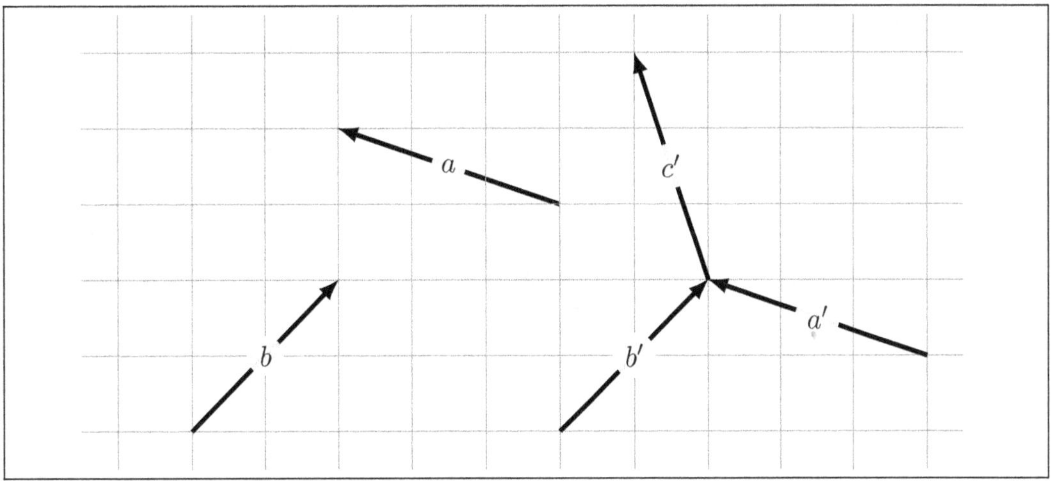

Figure 4: Illustration for axiom AxInecoll

AxInecoll If $k \in \mathsf{IOb}$ and $a, b \in \mathsf{Ip}$ such that $v_k(a) < \infty$, $v_k(b) < \infty$ and $m_k(a) + m_k(b) \neq 0$, then there are $a', b' \in \mathsf{Ip}$ such that a' and b' collide inelastically, with $m_k(a') = m_k(a)$, $\mathbf{v}_k(a') = \mathbf{v}_k(a)$, $m_k(b') = m_k(b)$ and $\mathbf{v}_k(b') = \mathbf{v}_k(b)$.[4] See Fig. 4.

4 Formulating the thought experiments

Here we are going to formalize and prove the thought experiments of Subsections 2.1, 2.2 and 2.3.

Formula ∃NegMass below says that there is at least one inertial particle of finite speed and negative relativistic mass.

∃NegMass There are $k \in \mathsf{IOb}$ and $a \in \mathsf{Ip}$ such that $m_k(a) < 0$ and $v_k(a) < \infty$.

Formula ∃FTLIp below says that there is at least one faster than light inertial particle.

∃FTLIp There are $k \in \mathsf{IOb}$ and $b \in \mathsf{Ip}$ such that $1 < v_k(b) < \infty$.

[4]Because here we use the framework of [3], we express possible worldlines of particles using existential quantifiers as is usual in frameworks of classical logic. See [25] for an axiomatic framework where this possibility is expressed instead by a modal logic operator.

4.1 First thought experiment

Axiom $\mathsf{AxThExp_1}$ below says that the thought experiment described in Subsection 2.1 can be done by asserting that inertial observers can send out particles moving with the speed of light 1 in any direction and having arbitrary positive relativistic mass.

$\mathsf{AxThExp_1}$ For $k \in \mathsf{IOb}$, $m \in Q$ and $\mathbf{v} \in Q^3$ for which $m > 0$ and $|\mathbf{v}| = 1$, there is $b \in \mathsf{Ip}$ such that $\mathbf{v}_k(b) = \mathbf{v}$ and $m_k(b) = m$.

Proposition 1. *Assume* $\mathsf{ConsFourMomentum}$, $\mathsf{AxEField}$, AxIp, $\mathsf{AxInecoll}$, $\mathsf{AxThExp_1}$. *Then*

$$\exists\mathsf{NegMass} \ \Rightarrow \ \exists\mathsf{FTLIp}. \tag{3}$$

Proof. By axiom $\exists\mathsf{NegMass}$, there is an inertial observer k and inertial particle a such that $m_k(a) < 0$ and $v_k(a) < \infty$. Let $\mathbf{v} \in Q^3$ for which $|\mathbf{v}| = 1$. Then by axiom $\mathsf{AxThExp_1}$, there is an inertial particle b such that $m_k(b) = -2m_k(a)$ and

$$\mathbf{v}_k(b) = \begin{cases} \mathbf{v} & \text{if } v_k(a) = 0, \\ \frac{-\mathbf{v}_k(a)}{v_k(a)} & \text{if } v_k(a) \neq 0. \end{cases}$$

By axiom $\mathsf{AxInecoll}$, there are inelastically colliding inertial particles a', b' and c' such that $\mathsf{inecoll}_k(a'b' : c')$, $m_k(a') = m_k(a)$, $\mathbf{v}_k(a') = \mathbf{v}_k(a)$, $m_k(b') = m_k(b)$ and $\mathbf{v}_k(b') = \mathbf{v}_k(b)$. By $\mathsf{ConsFourMomentum}$,

$$\begin{aligned} m_k(c') &= m_k(a') + m_k(b') \\ &= m_k(a) + m_k(b) = -m_k(a) \end{aligned} \tag{4}$$

and

$$m_k(c')\mathbf{v}_k(c') = \begin{cases} -2m_k(a)\mathbf{v} & \text{if } v_k(a) = 0, \\ m_k(a)\mathbf{v}_k(a) + 2m_k(a)\frac{\mathbf{v}_k(a)}{v_k(a)} & \text{if } v_k(a) \neq 0. \end{cases} \tag{5}$$

Hence

$$\mathbf{v}_k(c') = \begin{cases} 2\mathbf{v} & \text{if } v_k(a) = 0, \\ -(v_k(a) + 2)\frac{\mathbf{v}_k(a)}{v_k(a)} & \text{if } v_k(a) \neq 0. \end{cases} \tag{6}$$

Therefore, $v_k(c') = |\mathbf{v}_k(c')| > 1$ and $v_k(c') < \infty$; and this is what we wanted to prove. □

4.2 Second thought experiment

Axiom $\mathsf{AxThExp_2}$ below ensures the existence of the particle having positive relativistic mass used in the thought experiment described in Subsection 2.2.

$\underline{\mathsf{AxThExp_2}}$ For every $k \in \mathsf{IOb}$ and $m > 0$, there is $b \in \mathsf{Ip}$ such that $v_k(b) = 0$ and $m_k(b) = m$.

Formula $\exists\mathsf{MovNegMass}$ below asserts that there is at least one moving inertial particle of finite speed and negative relativistic mass.

$\underline{\exists\mathsf{MovNegMass}}$ There are $k \in \mathsf{IOb}$ and $b \in \mathsf{Ip}$ such that $m_k(b) < 0$ and $0 < v_k(b) < \infty$.

For the sake of economy, we use axiom $\exists\mathsf{MovNegMass}$ instead of $\exists\mathsf{NegMass}$ because in this case we do not have to assume anything about the possible motions of inertial observers or the transformations between their worldviews. We note, however, that these two axioms are clearly equivalent in both Newtonian and relativistic kinematics (assuming that inertial observers can move with respect to each other).

Proposition 2. *Assume* ConsFourMomentum, AxEField, AxIp, AxInecoll, $\mathsf{AxThExp_2}$. *Then*

$$\exists\mathsf{MovNegMass} \Rightarrow \exists\mathsf{FTLIp}. \tag{7}$$

Proof. By axiom $\exists\mathsf{MovNegMass}$, there is an inertial observer k and inertial particle a such that $m_k(a) < 0$ and $0 < v_k(a) < \infty$. By axiom $\mathsf{AxThExp_2}$, there is an inertial particle b such that $m_k(b) = -m_k(a)\,(1 + v_k(a)/2)$ and $v_k(b) = 0$. By axiom AxInecoll, there are inelastically colliding inertial particles a', b' and c' such that $\mathsf{inecoll}_k(a'b'\!:\!c')$, $m_k(a') = m_k(a)$, $\mathbf{v}_k(a') = \mathbf{v}_k(a)$, $m_k(b') = m_k(b)$ and $\mathbf{v}_k(b') = \mathbf{v}_k(b)$. By ConsFourMomentum,

$$\begin{aligned} m_k(c') &= m_k(a') + m_k(b') \\ &= m_k(a) + m_k(b) = \frac{-m_k(a)v_k(a)}{2} \end{aligned} \tag{8}$$

and

$$m_k(c')\mathbf{v}_k(c') = m_k(a)\mathbf{v}_k(a). \tag{9}$$

It follows that

$$\mathbf{v}_k(c') = -2\frac{\mathbf{v}_k(a)}{v_k(a)},$$

and hence that $v_k(c') = 2 > 1$, which is what we wanted to prove. □

4.3 Third thought experiment

Finally let us introduce the following axiom ensuring the existence of the particles having positive relativistic mass needed in the thought experiment of Subsection 2.3.

AxThExp3 For all $\varepsilon > 0$, $k \in \mathsf{IOb}$ and $a \in \mathsf{Ip}$, there is $b \in \mathsf{Ip}$ such that $(1+\varepsilon)|m_k(a)| < m_k(b) < (1+2\varepsilon)|m_k(a)|$ and $\mathbf{v}_k(a) = -\mathbf{v}_k(b)$.

Proposition 3. *Assume* ConsFourMomentum, AxEField, AxIp, AxInecoll, AxThExp3. *Then*

$$\exists \mathsf{MovNegMass} \Rightarrow \exists \mathsf{FTLIp}. \tag{10}$$

Proof. By axiom ∃MovNegMass, there is an inertial observer k and inertial particle a such that $m_k(a) < 0$ and $0 < v_k(a) < \infty$. Let $0 < \varepsilon < v_k(a)$. Then by axiom AxThExp3, there is an inertial particle b such that $(1+\varepsilon)|m_k(a)| < m_k(b) < (1+2\varepsilon)|m_k(a)|$ and $\mathbf{v}_k(b) = -\mathbf{v}_k(a)$.

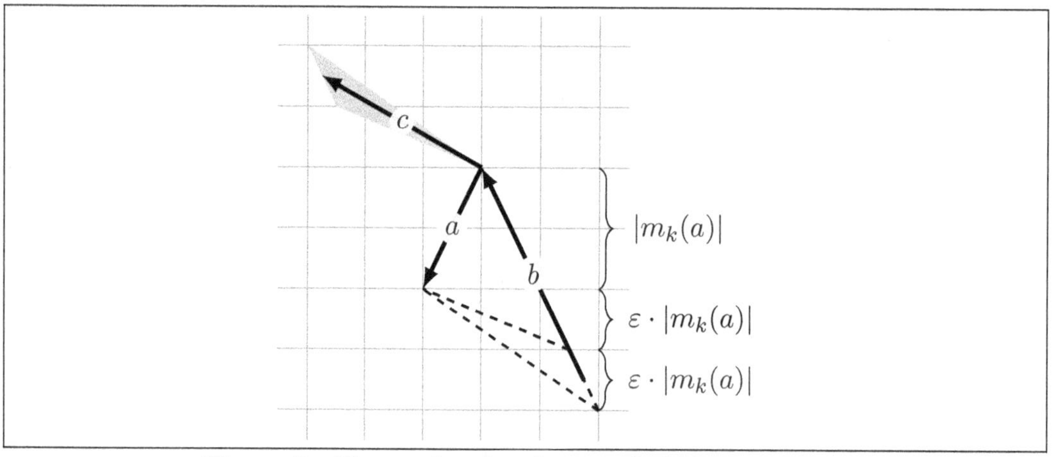

Figure 5: Illustration for the proof of Proposition 3

By axiom AxInecoll, there are inelastically colliding inertial particles a', b' and c' such that $\mathsf{inecoll}_k(a'b' : c')$, $m_k(a') = m_k(a)$, $\mathbf{v}_k(a') = \mathbf{v}_k(a)$, $m_k(b') = m_k(b)$ and $\mathbf{v}_k(b') = \mathbf{v}_k(b)$. By ConsFourMomentum,

$$\varepsilon|m_k(a)| < |m_k(c')| < 2\varepsilon|m_k(a)| \tag{11}$$

and

$$2|m_k(a)|v_k(a) < (2+\varepsilon)|m_k(a)|v_k(a) < |m_k(c')\mathbf{v}_k(c')|. \tag{12}$$

Hence

$$v_k(c') = |\mathbf{v}_k(c')| > \frac{2|m_k(a)|v_k(a)}{2\varepsilon|m_k(a)|} > \frac{v_k(a)}{\varepsilon}. \tag{13}$$

Therefore, $1 < v_k(c') < \infty$; and this is what we wanted to prove. □

5 Concluding remarks

Using only basic postulates concerning the conservation of four-momentum, we have shown axiomatically that the existence of particles having negative relativistic masses implies the existence of FTL particles. The following are the two most straightforward applications of this result.

- If an experiment eventually shows the existence of particles having negative masses, then we will know that FTL particles must also exist. If evidence exists suggesting otherwise, our approach would then imply that one or more of the natural assumptions encoded in our axioms must be false. This in turn would provide information suitable for guiding further experimentation.

- Similarly, if we can prove that FTL particles cannot exist, and *no* evidence can be found suggesting that the natural physical assumptions encoded by our axioms are invalid, then this can be used to prove the non-existence of particles having negative masses.

It is also worth noting that we have made no restrictions on the worldview transformations between inertial observers. Hence our axioms are so general that they are compatible with both Newtonian and relativistic kinematics. In addition to making our axioms relatively easy for students to understand, and hence our results more believable, the benefit of being so parsimonious with the basic assumptions is that it makes results obtained using our axiomatic method that much more difficult to challenge, because so few basic assumptions have been made concerning physical behaviours in the "real world".

Acknowledgements

This research was partially supported by the Royal Society [grant number IE110369]; London Mathematical Society [grant number 41508]; and by the Hungarian Scientific Research Fund [grant numbers T81188, PD84093].

References

[1] H. Andréka, J. X. Madarász, and I. Németi. Logic of space-time and relativity theory. In M. Aiello, I. Pratt-Hartmann, and J. van Benthem, editors, *Handbook of spatial logics*, pages 607–711. Springer-Verlag, Dordrecht, 2007.

[2] H. Andréka, J. X. Madarász, I. Németi, M. Stannett, and G. Székely. Faster than light motion does not imply time travel. *Classical and Quantum Gravity*, 31(9):095005, 2014.

[3] H. Andréka, J. X. Madarász, I. Németi, and G. Székely. Axiomatizing relativistic dynamics without conservation postulates. *Studia Logica*, 89(2):163–186, 2008.

[4] H. Andréka, J. X. Madarász, I. Németi, and G. Székely. On logical analysis of relativity theories. *Hungarian Philosophical Review*, 2010/4:p.204–222, 2011.

[5] H. Andréka and I. Németi. Comparing theories: the dynamics of changing vocabulary. In *Johan V. A. K. van Benthem on logical and informational dynamic*, pages 143–172. Springer Verlag, 2014.

[6] Hajnal Andréka, Judit X. Madarász, and István Németi. Logical axiomatizations of space-time. samples from the literature. In András Prékopa and Emil Molnár, editors, *Non-Euclidean Geometries: János Bolyai Memorial Volume*, pages 155–185, Boston, MA, 2006. Springer US.

[7] T. W. Barrett and H. Halvorson. Glymour and Quine on theoretical equivalence. *Journal of Philosophical Logic*, 45(5):467–483, 2016.

[8] T.W. Barrett and H. Halvorson. Morita equivalence. *The Review of Symbolic Logic*, 9(3):556–582, 2016.

[9] J. Belletête and M. B. Paranjape. On negative mass. *International Journal of Modern Physics D*, 22:1341017, 2013.

[10] H. Bondi. Negative Mass in General Relativity. *Reviews of Modern Physics*, 29(3):423–428, July 1957.

[11] C. J. Dias and L. N. Gonçalves. The concept of negative mass and its application to Helium-balloon-type dynamics. *American Journal of Physics*, 82(10):997–1000, 2014.

[12] A. Einstein. Zur Elektrodynamik bewegter Körper. *Annalen der Physik*, 17:891–921, 1905.

[13] A. Föppl. Über eine Erweiterung des Gravitationsgesetzes. *Sitzungsber. Kgl. Bayr. Akad. Wiss.*, 6:93, 1897.

[14] M. Friend. On the epistemological significance of the Hungarian project. *Synthese*, 192(7):2035–2051, 2015.

[15] N. S. Govindarajulu, S. Bringsjord, and J. Taylor. Proof verification and proof discovery for relativity. *Synthese*, 192(7):2077–2094, 2015.

[16] R. T. Hammond. Negative mass. *European Journal of Physics*, 36(2):025005, 2015.

[17] M. Hohmann and M. N. R. Wohlfarth. No-go theorem for bimetric gravity with positive and negative mass. *Physical Review D*, 80:104011, 2009.

[18] S. Hossenfelder. Bimetric theory with exchange symmetry. *Phys. Rev. D*, 78:044015, 2008.

[19] S. Hossenfelder. Comment on "No-go theorem for bimetric gravity with positive and negative mass". arXiv:0909.2094, 2009.

[20] M. Jammer. *Concepts of Mass in Contemporary Physics and Philosophy.* Princeton University Press, Princeton, NJ, 2000. pp. 129–130.

[21] J. X. Madarász, M. Stannett, and G. Székely. Why do the Masses and Momenta of Superluminal Particles Decrease as their Velocities Increase. *SIGMA*, 10:21, 2014. arXiv:1309.3713 [gr-qc].

[22] J. X. Madarász and G. Székely. Special relativity over the field of rational numbers. *International Journal of Theoretical Physics*, 52(5):1706–1718, 2013.

[23] J. X. Madarász and G. Székely. The existence of superluminal particles is consistent with relativistic dynamics. *Journal of Applied Logic*, 12(4):477–500, 2014.

[24] S. L. Meyer. Another example of the use of the "negative mass" concept in mechanics. *American Journal of Physics*, 38(12):1476–1477, 1970.

[25] A. Molnár and G. Székely. Axiomatizing Relativistic Dynamics using Formal Thought Experiments. *Synthese*, 192(7):2183–2222, July 2015.

[26] J. Noldus and P. Van Esch. Rebuttal on 'anti-gravitation' by S. Hossenfelder. *Phys. Lett. B*, 639:667–669, 2006.

[27] R. H. Price. Negative mass can be positively amusing. *American Journal of Physics*, 61(3):216–217, 1993.

[28] S. Rosenstock, T. W. Barrett, and J. O. Weatherall. On Einstein algebras and relativistic spacetimes. *Studies in History and Philosophy of Science Part B: Studies in History and Philosophy of Modern Physics*, 52, Part B:309–316, 2015.

[29] M. Stannett and I. Németi. Using Isabelle/HOL to Verify First-Order Relativity Theory. *Journal of Automated Reasoning*, 52:361–378, 2014.

[30] G. Székely. The existence of superluminal particles is consistent with the kinematics of Einstein's special theory of relativity. *Reports on Mathematical Physics*, 72(2):133–152, 2013.

[31] J. O. Weatherall. Are Newtonian gravitation and geometrized Newtonian gravitation theoretically equivalent? *Erkenntnis*, 81(5):1073–1091, 2016.

[32] M. Wenzel. The Isabelle/Isar Reference Manual. `http://isabelle.in.tum.de/dist/Isabelle2012/doc/isar-ref.pdf`, 2012.

[33] S. Ziauddin. Concept of "negative mass" in solving a problem in freshman statics. *American Journal of Physics*, 38(3):384–385, 1970.

Received 13 June 2017

Estimating the Strength of Defeasible Arguments: A Formal Inquiry

James B. Freeman
Hunter College of the City University of New York, New York, NY 10065, USA.
jfreeman@hunter.cuny.edu

Abstract

We propose that strength for defeasible arguments be understood as resistance to rebuttals: the greater the resistance, the stronger the argument. We may explicate this characterization through L. J. Cohen's method of relevant variables. A relevant variable is a condition which may hold in different ways in different situations and may hold more or less, and not just all or none. In some cases, if a variable holds in some way or to some extent, some universal generalization, most simply of the form that all *P*s are *Q*s, may be counterexampled. Likewise the corresponding warrant from Px to infer Qx will be rebutted by this condition. The more such variants of such variables do not produce counterexamples or rebuttals, the stronger the generalization and its associated warrant and the stronger the argument. A canonical test systematically exposes a generalization to progressively greater combinations of relevant variables. The more levels passed without counterexample, the stronger the generalization. Our strategy for explicating argument strength requires defining the concept of a relevant variable and indicating a canonical way to order relevant variables in constructing a canonical test. After explicating Cohen's concept, we develop how relevant variables may be ordered through appealing to the concept of plausibility and Rescher's account of plausibility indexing.

Keywords: canonical test, counterexample, L. J. Cohen, method of relevant variables, plausibility, plausibility indexing, rebuttal, N. Rescher, universal generalization, warrant

I want to thank the two anonymous referees of IfCoLog for their very helpful comments on a previous version of this paper, which I believe have greatly improved it.

1 Introduction

Consider the following argument:

1. Martina will do well in college, because
2. She scored high on the Scholastic Aptitude Test and
3. She has demonstrated high scholastic motivation.

Intuitively, we want to count this argument stronger than the arguments from (2) or (3) alone as premises to (1) as conclusion. But why? Before proceeding further, it will be useful to review certain aspects of the Toulmin model (See [10, pp. 97–107]), his view of the "layout of arguments." Toulmin begins by distinguishing claims, data, and warrants. A claim is what an argument seeks to establish or support. Data are facts–using "fact" very broadly–put forward to support a claim. A warrant answers the question of how one gets from the data to the claim. It is the principle of reasoning used in the argument, an inference licence.

Consider the warrant of the argument from (2) and (3) to (1):

From: x scored high on the Scholastic Aptitude Test and x has demonstrated high scholastic motivation

To infer: x will do well in college[1]

The argument from this conjunction is not subject to a rebutting defeater the way an argument from just (2) or (3) to (1) is. Within the framework of formal disputation [8, pp. 1–24], we can model the genesis of the argument from (2) to (1) this way:

Proponent	Challenger
(1) Martina will do well in college.	Please show that Martina will do well in college.
(2) Martina scored high on the Scholastic Aptitude Test.	

The challenger can obviously attack the proponent's argument by raising the issue of a rebutting defeater, a rebuttal for short, a condition logically consistent with the premise but negatively relevant to the conclusion, such as Martina's having low scholastic motivation. She can extend the exchange this way:

[1]Notice that we identify the warrant as licencing the move from the conjunction of the premises to the conclusion. We are not reading this argument as having two warrants, one from (2) to (1) and the other from (3) to (1), their separate strength to be somehow arithmetically combined to determine the overall strength of the argument.

(2) Martina scored high on the Scholastic Aptitude Test.	*Ceteris paribus* Martina will not do well in college if she does not have high scholastic motivation. But for all you have shown, she does not have high scholastic motivation, i.e. please show that she does.

The proponent's response to the challenger's second move may very well be to add the second premise above, i.e. (3). Clearly, the challenger could not have introduced her rebuttal to the move from (2) and (3) conjointly to (1) because it would be inconsistent with the second conjunct altogether. If the proponent had presented the conjunction of (2) and (3) as the initial move to his argument, he would have made a pre-emptive strike against the challenger's attack at step (2). The rebuttal is already countered. The moral of the story, as we see it, is that the warrant of the argument from (2) and (3) to (1) is more rebuttal-resistant than the warrant from the argument just from (2) alone to (1) or from (3) alone to (1). We propose then that comparative argument strength (for defeasible arguments) be understood as resistance to rebuttals: the more resistant to rebuttal, the stronger the argument.

2 The Method of Relevant Variables

We further propose that we may refine this intuitive thesis and frame an argument for it by developing L. J. Cohen's account of the method of relevant variables in [2, 3, 4]. Suppose we observe a correlation between occurrences of m-ary properties P and Q. In accordance with Cohen's conception, let us assume the correlation is universal rather than statistical. Our observations then back the warrant

From: $\mathbf{P}x_1, \ldots, x_m$

To infer: $\mathbf{Q}x_1, \ldots, x_m$

We have made our observations in a default condition. There may be one or more other factors, operating in the current case, needed for something's being a **Q** even when it is **P**. Likewise other factors preventing its being a **Q** are not operating in this case. We have made no attempt to investigate whether varying the degree to which these other factors are present affects whether there may be **P**s which are not **Q**. These other factors are variables and the degree to which they are present are variants of that variable. Should an instance of **P** which is also an instance of a variant **v** of some variable **V** not be **Q**, we have a rebuttal to our warrant. Likewise, if we were to take

$$(\forall x_1)\ldots(\forall x_m)(\mathbf{P}x_1,\ldots,x_m \supset \mathbf{Q}x_1,\ldots,x_m)$$

as the associated generalization of the warrant, we would have a counterexample to that generalization.

Since the cases in our default situation show constant conjunction, but no variables have been varied, we say that our warrant is backed to degree 1 or alternatively that our universal generalization has received first level of support, support to degree $1/n$, where n is the number of recognized relevant variables. In the method of relevant variables, we identify and order a finite number n of these variables. We then conduct a canonical test. At degree one, we consider just the default situation, where we vary only the values of the first variable. Finding no counterexample at this level, we proceed to level two of the test, where we consider variants of the second relevant variable singly and in combination with values of the first variable. If no counterexample appears, the warrant and its associated generalization pass level two. The test then continues until some level i+1 reveals a counterexample or we proceed all the way through to level n with no counterexample appearing. A counterexample at $i+1$ but at no $j \leq i$ allows us to say that the generalization is supported to degree i/n. No counterexample appearing at any level of the test constitutes n/n level of support and, in Cohen's view, identifies the generalization as a law of nature. This is not to say that the generalization is a logically necessary statement or that the evidence gathered by the canonical test necessitates the generalization. There may be unrecognized relevant variables with variants which do constitute counterexamples. There is no logical impossibility here. But until and unless such a relevant variable appears, we are justified in reasoning according to our warrant without qualification or hedging. We are proposing then that we may estimate argument strength through degree of support by a canonical test. Where $i > j$, a level of support to i/n is greater than a level of support to j/n.

Our account of the method of relevant variables is seriously incomplete on two grounds. First, how does one define or identify a relevant variable? Second, how does one order the set of relevant variables once they are identified? Cohen has addressed the first problem, which we address in the next section. The ordering question will occupy the remaining sections of this paper.

3 The Problem of Defining Relevant Variables

We must understand a relevant variable relative to a universal generalization and, following Cohen, should understand it relative to a set of universal generalizations. Although his characterization of a relevant variable is complex, natural kinds may furnish a straightforward motivation. Giraffes and horses are distinct natural kinds and there is no overlap between them. Clearly, without worrying about biological issues, we may regard giraffes and certain other natural kinds as all species within a given genus. Where $\mathbf{s}_1,\ldots,\mathbf{s}_k$ are the species within a genus $\mathbf{G}, S_i, 1 \leq i \leq k$ indicate the predicates saying that an element $\mathbf{e}$ is a member of species $\mathbf{s}_i$. We refer to these predicates as species predicables. In addition, there will be a distinct set of predicables, mutually exclusive of the $S_i, T_j, 1 \leq j \leq k$. Following Cohen, we may call these the target predicables. Observation may indicate that for some species predicables and target predicable, universal generalizations hold of that species, i.e. observation confirms '$(\forall x)(S_i x_1,\ldots,x_n \supset T_j x_1,\ldots,x_n)$' holds. In the simplest case, where S_i, T_j are monadic predicates, we have that observation confirms that '$(\forall x)(S_i x \supset T_j x)$' holds. For simplicity, let us consider just monadic predicates here.

Consider just three natural kinds of animals, horses, giraffes, and zebras. For the purposes of our discussion here, let us call these three kinds species and let S_h, S_g, S_z be the three species predicables. Suppose we want to test the extent to which some generalization '$(\forall x)(Px \supset Qx)$' holds of giraffes. We have observed a constant correlation. Could that correlation be counterexampled? Let $V_1,\ldots,V_5$ be properties which can hold of giraffes, horses, and zebras to some degree. Consider V_1. Let v_1^i indicate a particular degree of V_1. Where e is a horse, suppose observation shows that '$(Pe \,\&\, v_1^i e) \,\&\, \sim Qe$' holds, i.e. e is a counterexample to '$(\forall x)(Px \supset Qx)$' for horses. Given the analogy of horses, giraffes, and zebras, let us count the union of the sets of horses, giraffes, and zebras as a genus containing these three species. Thus, it is possible that where 'h' indicates a horse, $v_1^i h$ constitutes a counterexample to '$(\forall x)(Px \supset Qx)$' for horses, as may '$v_1^i h$' for other degrees of V_1. That is, observation has shown that V_1 is a potential relevant variable for the class of horses when the issue is whether all horses which are Ps are also Qs. We understand our relevant variables to be relevant variables for that genus. The observation of the failure of a universal generalization for horses when some value of a relevant variable also holds raises the question of whether the generalization will fail for other species within that genus, in particular giraffes, even if the results of no test of whether the generalization for giraffes holds, despite values of the relevant variable also holding, have been observed.

Consider some natural kind, e.g. tiger, wasp, tulip, gold. Observation may show that certain generalizations hold constantly or universally for the members of this natural kind. For example, restricting our universe of discourse just to tigers, "Whenever offered meat, it will eat the meat," '$(\forall x)(Ox \supset Ex)$'. The question now arises of whether this is simply an accidental correlation or whether there is a more law-like connection for tigers between being offered meat and eating the meat offered. The class of tigers is a subset of a wider class, if anything the class of felines. Let us suppose, for sake of illustration, that some members of that wider class will refuse meat under certain circumstances, i.e. '$(Ox \ \& \ Rx) \ \& \sim Ex$' holds of those members of that class. 'Rx' abbreviates some further property, which when holding of some feline, the animal instances a counterexamle to '$(\forall x)(Ox \supset Ex)$.' Likewise '$Rx$' constitutes a rebuttal to the inference from 'Ox' to 'Ex.' As lions, cougars, pumas, house cats are subclasses of felines, so observation may show that some members of some of these classes under some conditions instance courterexamples to '$(\forall x)(Ox \supset Ex)$.' Let $V_1, ..., V_n$ list these counter-exampling properties. We may assume that at least some of the V_i come in degrees. The degree $\mathbf{v}_{i-j}$ constitutes a variant of the variable V_i. $V_1, \ldots, V_n$ then are the relevant variables for the class of felines (at least those relevant variables which are known for that class). The inquiry via the method of relevant variables on whether being offered meat and eating the meat offered are more connected in a law-like way than what an observation of constant co-variation among tigers shows consists of offering meat to one or more tigers which satisfy some degree of a recognized relevant variable V_i, recognized through observation of members of other species in the genus who refuse meat when offered and who also satisfy a degree of some V_i. If the tigers all satisfying some degree of the relevant variable all eat the meat offered, then our canonical test has shown the generalization "Whenever offered meat, the tiger will eat the meat," has passed the ith level of the test. Of course, the generalization may fail for other relevant variables which have yielded counterexamples for other members of the genus. However, the fewer the relevant variables which yield counterexamles when tested on tigers, the tighter the connection between 'Ox' and 'Ex' for tigers.

It is important to emphasize two points here.[2] First, recognizing that when members of some species, other than tigers but within the same genus, satisfy some property P (to some degree) in addition to O, they fail to satisfy E, we have identified P as a potential or possible relevant variable for tigers. If there have been no previous tests of the generalization '$(\forall x)(Ox \supset Ex)$' for tigers, until we run a canonical test on tigers which satisfy 'O' and 'P' to some degree and find failure to satisfy 'E', we shall not have found that 'P' or degrees of 'P' yield counterexamples to

[2]We wish to thank an anonymous referee of *IfCoLog* for pointing out to us the objections these two points seek to meet.

'$(\forall x)(Ox \supset Ex)$.' The point is that from V's being a relevant variable for some genus, we cannot infer that values of V will generate counterexamples for all the species within the genus.

The second point to emphasize is that where 'P^i' expresses the property of satisfying the ith degree of the relevant variable P, 'P^i' and 'O' must be simultaneously satisfiable, i.e. we can consistently predicate 'P^i' and 'O' of any element in the species over which we are testing '$(\forall x)(Ox \supset Ex)$.' Recognizing this point allows us to meet an objection to our understanding of 'rebuttal' and 'relevant variable.' Suppose we have included in our genus of which tiger is one species, a species which cannot digest meat and because of this the members of the species will refuse meat when offered. Would values of the variable 'not being able to digest meat' (surely not being able to digest meat may come in degrees) constitute potential rebuttals to the generalization that tigers offered meat eat the meat. Should we count 'not being able to digest meat' a relevant variable when setting up a canonical test for meat eating for tigers? I believe the answer is no. Although the property of being a tiger may be *logically* compatible with being unable to digest meat, one can argue that not being able to digest meat is *essentially* incompatible with being a tiger .It would certainly be a very unusual tiger who could not digest meat. In [6], Kornblith points out that objects belonging to a natural kind (or in some cases an artificial kind) have and can be perceived to have "insides" and "outsides." "Outsides" involve superficial properties of an object, such as color or–for certain animals–furriness. These properties are open to inspection through direct perception. "Insides" by contrast deal with causal factors which may explain the outside or surface properties and which do determine whether an object is a member of that kind. If one does not object to speaking of essences, the "insides" constitute the essence of the object. (See [6, Chapter 4 and especially 4.3].) Hence, lacking an essential property is an indication that an object is not a member of that natural kind.

Properties such as mode of nutrition and mode of reproduction are essential properties of a natural kind of living thing. So is a tiger who cannot digest meat really a tiger? Obviously we cannot entertain an exploration of that issue here. What we can say is that if we count being able to digest meat as of the essence or consequent upon the essence of being a tiger, then not being able to digest meat should not be counted among the relevant variables in setting up a canonical test of '$(\forall x)(Ox \supset Ex)$,' for tigers since the property is incompatible with being a tiger. We are dealing with an essential, if not logical, incompatibility. On the other hand, if being able to digest meat is not of the essence of being a tiger, we still need not include not being able to digest meat (to some degree) as a relevant variable in setting up a canonical test, if as a matter of fact it is known that there are no tigers which cannot digest meat. If 'not being able to digest meat' were included as a relevant variable in a canonical test, our generalization would pass this level vacuously.

We may make a similar reply to a second issue. Consider the warrant

From: x is a tiger & x is more than 2.5 years old & x has not given birth in the past 2.5 years

To infer: x lives alone

We may accept that tigers which are more than 2.5 years old live alone except if they have given birth within the past 2.5 years. But lions, which presumably will be included in the same genus as tigers, may not live alone. Let us assume that not being able to get an adequate amount of food if living alone explains why lions do not live alone. So does 'not being able to get sufficient food when living alone' constitute a potential rebuttal to our warrant? Again, I believe the answer is no. Surely there is more than an accidental connection between being a tiger and being able to get enough food even if living alone, at least when more than 2.5 years old. Tigers who are more than 2.5 years old but not able to get sufficient food are somehow defective in what is essential to being a tiger. So either one would not include this rebuttal as a relevant variable in a canonical test or a canonical test would pass the level of this relevant variable vacuously. Having given this motivation, we can now turn to Cohen's general formal characterization of a relevant variable.

Assume $\mathbf{P}_1,\ldots,\mathbf{P}_k$ are j-ary predicate expressions whose extensions are mutually exclusive classes, e.g. distinct natural kinds. $\mathbf{Q}_1,\ldots,\mathbf{Q}_m$, are further predicates which may be true of the elements in the extension of $\mathbf{P}_1,\ldots,\mathbf{P}_k$, e.g. 'is a herbivore,' 'bears its young live.' The sets $\{\mathbf{P}_1,\ldots,\mathbf{P}_k\},\{\mathbf{Q}_1,\ldots,\mathbf{Q}_m\}$, have no members in common. The P_is are j-ary predicates to the effect that the j-ary sequences satisfying these predicates are instances of some "natural kind." The Q_is are predicates that some j-ary attribute holds of some j-ary sequence. It must be semantically meaningful to predicate these attributes of sequences satisfying some natural kind predicable. Notice that $\mathbf{P}_1,\ldots,\mathbf{P}_k$ need not be atomic predicates. In particular, we may form conjunctions of these predicates as long as we use a different individual variable with each conjunct. Since the P_is, $1 \leq i \leq k$ are mutually incompatible, using the same variable would result in a predicate with empty extension, e.g. nothing is both a horse and a giraffe. But relations may hold constantly between the members of two kinds, e.g. 'having a longer neck than.' Hence for example, the sentence 'Giraffes have naturally longer necks than horses' becomes

$$(\forall x)\forall y)([Gx\&Hy] \supset Lxy),$$

a sentence perfectly meaningful and not vacuously true.

Suppose that observation has shown that members of a given natural kind always satisfy some property or relation. For example, observation may show that giraffes are especially fond of the leaves of a given tree. Suppose chemical analysis shows that these leaves are especially rich in a given nutrient. Does the abundance of this nutrient explain giraffe preference or is some other factor or combination of factors present in the leaf causing the behavior? After all, the leaves are, for giraffes, conveniently located, they are all green, and they all contain various other nutrients. Now convenience of food source, color of food source, nutrients available in the source are all factors which for other kinds of herbivores may exert an established causal influence on whether they prefer that food source. Clearly color, convenience of access, and amount of nutrient present can vary by more or less. This factor motivates regarding different amounts of each as variants of a variable, but does not rule out that in some cases, the variants may simply be the presence or absence of the variable. So what are the giraffes responding to? The point is this. We are assuming that giraffes consistently prefer one type of leaf rich in a certain nutrient necessary to giraffe flourishing. So where "*Gx*' abbreviates '*x* is a giraffe,' '*Pxy*' abbreviates '*x* prefers *y*,' and '*Nx*" abbreviates '*x* is an available food source containing nutrient *n*,' observation supports

$$(\forall x)(\forall y)(Gx \supset [Ny \supset Pxy])$$

where '*x*' ranges over the class of giraffes. This body of observation gives the universal generalization basic or first-degree support.

This support may be tested experimentally by considering variants of the variables identified. The factors our observation has shown which affect whether other types of herbivores prefer a given food source satisfying that factor to some degree identify these factors as relevant variables. So to see whether our generalization has more than first-degree support for giraffes, we must test it when degrees of those other factors are present (i.e. variants of those other variables). One might place the leaves various distances from the ground. One might change the color of the leaves or their taste. One might both vary the distance of the leaves from the ground and alter the color or taste of the leaves, or both. That is, one might carry out a canonical test, having identified by observation of horses, deer, rabbits, and squirrels a set of relevant variables. Clearly, the more the variants of these variables fail to produce counterexamples to the generalization, the stronger the support of the generalization and the stronger the backing we have for the warrant

From:	Gx
To infer:	$(\forall y)(Ny \supset Pxy)$

However, as we have indicated, our account is not yet complete. Perhaps there is little reason to think that an instance of one relevant variable would constitute a counterexample, while there may be good reason to suspect that instances of another would. The weight of the evidence for the generalization and the strength of the backing for the warrant might vary significantly depending on the order in which the relevant variables were addressed in the canonical test. Is there a "canonical" way of ordering the variables in a canonical test? We begin that investigation in the next section.

4 The Problem of Ordering Relevant Variables

Central to ordering relevant variables, according to Cohen, is the concept of greater falsficatory potential. In empirical investigation, whether variable $\mathbf{V}_i$ is seen to have greater falsificatory potential than $\mathbf{V}_j$ reflects empirical observation and is subject to revision over time. The ordering also presupposes that the number of relevant variables, together with the number of variants within each relevant variable, is finite. The rationale for this assumption is that any postulate that a given finite set of relevant variables or given finite set of variants within a variable is exhaustive is open to empirical refutation, while postulates of infinite variation are not. According to Cohen, "In practice, within a particular field of enquiry, experimental scientists generally have a rough scale of importance for relevant variables, depending partly on the accepted falsificatory efficiency of their variants and partly on convention" [4, p. 148]. Available empirical evidence then contributes to this accepted falsificatory efficiency. A set of relevant variables then will be ordered according to perceived lessening strength of falsificatory potential. What influences this perception? According to Cohen, "The greater the variety of types of hypotheses that a particular relevant variable is seen to falsify, the more important it will normally be presumed to be" [3, p. 141]. Here falsificatory potential parallels explanatory potential, a plausibility desideratum for explanatory hypotheses. *Ceteris paribus*, the hypothesis which explains more is more preferable. Cohen immediately adds that "considerations of simplicity, fruitfulness, technological utility, etc., often have to be taken into account besides considerations of evidential support" [3, p. 142], in hypothesizing the proper order of relevant variables. But simplicity, fruitfulness, utility are factors contributing to plausibility, as we shall see.

In setting up a canonical test, then, the relevant variables should be ordered according to the empirical information that we have concerning how likely they are to generate counterexamples to the generalization being tested. Suppose to adapt Cohen's example [3, pp. 129–133] one undertakes experiments to determine which colors, if any, bees may discriminate. Suppose that observation shows that when a colony of bees is trained to return to a blue card on which sugar water has been placed, bees continue to return to the card when there is no sugar water. What explains this behavior? One hypothesis is that the bees recognize the color blue. That bees are responding to the shape of the card, its relative position to the ground, its height off the ground, are alternative hypotheses involving relevant variables. Various particular shapes of the card, particular angles with respect to the ground, particular distances from the ground, are variants of these variables. In addition, bees as a natural kind subdivide into species. If only some species of bees discriminate colors, at least the color blue, that fact will not become evident just by observing one species of bee which does identify blue. So we have four variables. Which should we vary first in the canonical test?

5 Order Through Prior Probability

Bees as a natural kind are coordinate with certain other natural kinds–wasps, bumble bees, hornets to name just some. We may have some information antecedent to constructing and conducting our canonical test on bees about the sensitivity of insects of these other kinds to shape, position, height, and whether this sensitivity varies with respect to the species within a natural kind. If we regard our information about these other natural kinds as generating evidence for what relevant variables might affect bee behavior, we may regard our antecedent evidence concerning similar natural kinds as indicating the prior probability of some condition functioning as a relevant variable for bees. What we are asking for here is the prior probability for a particular condition that counterexamples to our generalization that all bees discriminate the color blue would be found by realizing determinates of that condition taken as a determinable. A proper answer to this question would both identify relevant variables and make ordering the variables a straightforward matter. Start with the variable having the highest prior probability of generating the most counterexamples and order the rest with decreasing prior probability. (Remember that our set of relevant variables is finite.) But how do we determine these prior probabilities?

We may have some information about how shape, relative position, height affects behavior of insects within a given class of insects appropriately analogous to bees at least with respect to visual mechanism. Is this information sufficient to justify a judgment of prior probability in general for canonical test purposes? Certain problems may obviously arise. We may very well have gaps in our information about types of cases. Evidence may show two relevant variables with the same number of counterexamples among their variants. How are these two to be ordered in particular on the grounds of prior probability? In connection with defending his view on confirmation theory, W. Salmon has discussed the notion of prior probability at length and connected it with the concept of plausibility, a concept which has been investigated by other philosophers, in particular N. Rescher. Can we construct an answer to our ordering question by consulting these accounts?

6 Prior Probability and Plausibility

6.1 A Standard Textbook Account of Plausibility

I. M. Copi and C. Cohen in [5] explicate the plausibility of a hypothesis through three properties: compatibility with previously established hypotheses, predictive or explanatory power, and simplicity. Barring new evidence coming to light which would call for a new hypothesis to explain that evidence together with the previously established evidence, and assuming that a previous hypothesis is well-confirmed, "the *presumption* is in favor of the older hypothesis" [5, p. 520], italics in original). Predictive or explanatory power refers to the body of facts deducible from or explained by a hypothesis. The greater this body of facts, the more powerful the theory. Finally, plausibility favors simplicity. A theory which identifies a single suspect as responsible for a crime is simpler than one which postulates a conspiracy. As Copi and Cohen point out, deciding which of rival hypotheses is simpler may in certain cases involve a judgment call subject to challenge. One hypothesis may be simpler than another in a given respect, while the reverse is true for a different respect. Which one, then, is simpler? May we link the concept of plausibility to prior probability to advance our understanding of how to order relevant variables according to their decreasing falsificatory efficiency? Let us turn to Salmon directly.

6.2 Salmon on Prior Probability and Plausibility

Salmon recognizes the plausibility of a hypothesis as involving "direct consideration of whether the hypothesis is of a type likely to be successful" [9, p. 118], i.e. direct consideration of its probability before taking into account a specific body of evidence. Relating this to the problem at hand, what is the probability that for $1 \leq i \leq 5, H_i$ (i.e. $\mathbf{V}_i$ produces more counterexamples to $(\forall x)(\mathbf{P}x \supset \mathbf{Q}x)$ than any $\mathbf{V}_j, j \neq i$) is true before ordering them in the design of a canonical test? Recall how one identifies a relevant variable. We are regarding Px as attributing having a certain property to n-tuples of a certain species **S**. But we recognize **S** as a natural kind among a class of natural kinds. Although we may not have near enough evidence about the **S**s to make any projection about any of the $\mathbf{V}_i$s with any acceptable degree of confidence, we may expect that we have more evidence when we take into account all the other species within the genus. This may still not be enough for a projection with confidence, but it may be all the information we have before designing and carrying out any canonical test. But this information should indicate which $\mathbf{V}_i$ produces more counterexamples to $(\forall x)(Px \supset Qx)$ than any other $\mathbf{V}_j$. On the basis of this data, however preliminary, we may rank the $\mathbf{V}_i, 1 \leq i \leq 5$. This ranking satisfies one of the major criteria for plausibility, i.e. compatibility with previously established results (allowing that these results may include data as well as theories). Why is this criterion truth-conducive? As Copi and Cohen explain, to yield reliable explanations (or generalizations), a theory must be supported by evidence. Hence a well-established theory can be relied on. Likewise, the more evidence one has that specific values of an attribute constitute counterexamples to a generalization, the more reliant one may be that those values will result in counterexamples in a new context.

Salmon indicates that another class of criteria may bear on plausibility ranking, which he calls pragmatic criteria. These criteria concern estimates of the reliability of the source of a claim. As such they relate directly to our example. We claim we have data based on observation of other species in the genus to which **S** belongs and which indicate which variable has the most variants which produce counterexamples to $(\forall x)(Px \supset Qx)$. What are the sources of this data and how trustworthy are these sources both in making observations and in reporting them sincerely? These considerations are also relevant to ranking the relevant variables for plausibility.

6.3 Rescher on Plausibility and Plausibility Indexing

Rescher is also concerned with these pragmatic criteria. In his discussion, he refers to the plausibility criteria of conformity to previous results, predictive or explanatory power, and simplicity as principles of inductive systematization [8, p. 41]. He indicates two more criteria: the authority or reliability of the source or sources vouching for a claim and "the probative strength of confirming evidence" [8, p. 35]. The word of a recognized authority renders some distinct plausibility to a claim, *ceteris paribus*. The scope of recognized authority extends beyond exclusively expert opinion to include any source which is "well-informed or otherwise in a position to make good claims to credibility" [8, p. 39]. So the word of someone in a position to have observed some event or a claim which has gained the status of common knowledge would count as plausible to some degree on this criterion [8, p. 39] and [7, p. 25].

What may we say of the sources which may vouch for a claim as markers of plausibility? By "source" Rescher means not just external sources which may vouch for a claim, such as testimony or common knowledge, but sources internal to the person entertaining the claim, such as sense perception. Including external and internal sources as factors bearing on the plausibility of claims indicates that the question of the plausibility for the claims whose ranking we are investigating depends not just on the quantity of evidence but its quality. To return to our example of five relevant variables, the claim that for $1 \leq i \leq 5$, $\mathbf{V}_i$ has among its variants more counterexamples to $(\forall x)(Px \supset Qx)$ than for any $\mathbf{V}_j, j \neq i$, gains plausibility from the quality of the sources vouching for the claims about the relevant variables themselves and about the individual counterexample variants. The question about the number of counterexamples concerns the specific variants which have produced counterexamples. Clearly, source **T** could vouch that $\mathbf{V}_i$ has more counterexamples among its variants than any other $\mathbf{V}_j$, but why should an interlocutor regard **T**'s word as reliable? So in ranking the plausibility of the claims about which relevant variable produces the most counterexamples, it is conceivable that more claims about particular variants of relevant variables of some $\mathbf{V}_j$ be recognized but the claim about $\mathbf{V}_i$ be regarded as the most plausible on the reliability, i.e. quality, of the sources vouching for it. Thus a conflict between ranking plausibility on the number of reported counterexamples produced by variants of a particular relevant variable and the quality of the sources reporting these counterexamples is possible.

Rescher has proposed a classification of the bases of plausibility into three groups, sources, confirming evidence, and principles of inductive systematization [8, p. 41]. The authoritativeness of the sources, the probative strength of the evidence, and the highest rank on principles of inductive strength are the contributing factors to the plausibility of a claim. We have seen that for our purposes, conformity with previous results and extent of probative evidence amount to the same thing. The statements whose plausibility we are trying to rank are claims that a particular relevant variable poses more counterexamples to a certain generalization than any other relevant variable. These statements are not universally generalized conditionals for which more specific laws or test conditionals might be derived. True, each statement yields the implication that in a test involving all the relevant variables, the variants of the relevant variable which is the subject of the statement will produce the more counterexamples. But this does not yield a preference for any of these statements over the others. Likewise these statements do not seem to differ in simplicity. Hence, in ranking the statements for plausibility, standing of sources in point of authoritativeness and probative strength of confirming evidence [8, p. 41], i.e. extent of analogous evidence, are the bases for ranking the statements for plausibility.

Suppose, as we have just indicated is possible, these sources conflict. Suppose the claim concerning V_i has the most confirming evidence but the claim concerning V_j is attested to by sources deemed more authoritative. How may we deal with this conflict and properly order V_i, V_j according to plausibility considerations? Rescher proposes grading sources according to their reliability. His scale parallels Cohen's scale for grading the legisimilitude of universal generalizations according to the method of relevant variables. We assume that there are n grades. The reliability of a source is graded by m/n, where $1 \leq m \leq n$. $n/n = 1$ "represents *maximal* or total reliability" [7, p. 7]. Only sources with at least some reliability need be considered. Those judged to have less than $1/n$ reliability "would be so unreliable that their data are effectively unusable" [7, p. 8]. If only sources of less than $1/n$ reliability have vouched for V_i, V_j, then they would not be ranked. But suppose at least one source of non-negligible reliability vouches that V_i contains more counterexamples than does any V_j and that there is conflict among sources possessing some positive degree of reliability. Clearly in this case, considering just questions of the reliability of sources, we rank statements according to the degree of reliability of the sources vouching for them. Ranking statements of the form "the number of counterexamples to $(\forall x)(Px \supset Qx)$ among the variants of V_i is greater than the number of counterexamples among the variants of V_j" for plausibility is now straightforward. We simply transfer the ranking of the sources to the ranking of the plausibility of the statement. If two or more sources of differing reliability vouch for the same statement, its plausibility will be the maximum of the reliability of the sources vouching for it. But we are still left with our principal question: How do we rank the reliability of sources which vouch for the claims about the relative falsificatory capacity of the relevant variables for some generalization $(\forall x)(Px \supset Qx)$? Have we not simply shifted the problem from ranking the plausibility of hypotheses to ranking the reliability of sources? At least for some, if not all of the sources vouching for a hypothesis, we shall have some reason to believe the source has some reliability. If we have no reason to believe that a source has at least minimal reliability, we can set that source aside. Such a source we judge to have less than $1/n$ reliability and thus regard it as so unreliable that its vouching for a claim does nothing to increase our confidence in that claim. This would be the case if someone simply vouched for the overall claim that among the variants of V_i are more counterexamples to $(\forall x)(Px \supset Qx)$ than among the variants of any other $V_j, 1 \leq i \leq 5$, where the person had no acquaintance–direct or indirect–with the capacity of the V_i to generate counterexamples to the claim.

How then may we rank reliability of sources? We are here considering sources which may vouch for two types of information–information about the number of counterexamples a relevant variable has produced across the "species" of some genus, and information about the compatibility of the hypothesis with previously established results. The trustworthiness of a source determines its reliability. I believe Rescher has addressed the issue of ranking source reliability in [8]. The reliability of a source is its "entitlement to qualify as well-informed or otherwise in a position to make good claims to credibility. It is on this basis that expert testimony and general agreement (the consensus of men [*sic*]) come to count as conditions for plausibility" [8, p. 41]. Let us elaborate. Suppose our knowledge of a source's cognitive background in some area indicated some positive level of competence or expertise in that area. Suppose also that we had no knowledge of any proclivity to dissemble or otherwise make false assertions on the part of that source. For example, should a trained and licenced oncologist present an interpretation of an x-ray which the oncologist had ordered, our recognition of this background and the absence of defeating evidence justifies our reposing a degree of confidence in the oncologist's word, indeed we would expect a relatively high degree of confidence. Again, suppose we had little information about the training or certification of the oncologist, but we did know that the physician had a very high track record of successful diagnoses, very few false positives or false negatives. Clearly, then, the oncologist has a proper entitlement to credibility. Again, suppose a number of sources which we recognize to possess a moderate amount of background in a subject area and a moderately good track record about rendering correct judgments in that area. We would be justified in according some degree of reliability to each source. Suppose all of the sources on a given occasion vouched for the same interpretive claim on the basis of the evidence. Rescher allows that we may count these sources which have converged on this point as one source, with higher reliability than any of the individual sources. This clearly stands to reason. Presumably these sources have diverse backgrounds, and thus the convergent source is rendering its opinion on a wider background in the subject area. In this way, we rank sources on the basis of our background knowledge of their reliability.

Let's apply these considerations to a case where hypotheses H_i, H_j conflict. On the basis of which of these two hypotheses is deemed most plausible, we shall decide which relevant variable should be tested first. We are considering the situation in which reports indicate that V_i produces more counterexamples to analogous generalizations concerning other species within the genus of the species about which we are generalizing. H_i is the claim that values of V_i produce the most counterexamples for the genus overall (and by analogy can be expected to produce the most counterexamples for $(\forall x)Px \supset Qx)$. However, past experience and theory suggest that V_j involves the most counterexamples for the genus and thus that H_j should be regarded as correct. The question then is which sources's word is to be deemed more authoritative. Rescher suggests [7, p. 111] that in different contexts different factors will determine the answer to this question. Since we are looking on both reports of observations and common knowledge or theoretical expectations, and seeking to adjudicate plausibility through witness reliability, the context for answering our question, according to Rescher, is information processing [7, p. 111]. What has Rescher said about information processing with inconsistent data which is germane to our question?

The first step in resolving the conflict posed by H_i versus H_j is to rate the reliability of the sources through an index from 0 to 1, with 1 high. A source deemed absolutely reliable gets an index of 1.0, relatively reliable 0.8 and somewhat reliable 0.5 [7, p. 75]. Where there is conflict, a source of higher reliability trumps those of lower reliability. Rescher makes this indexing of sources intuitively. So, to use his example, when the issue is reconstructing the text of a lost letter as accurately as possible, a photocopy, even fragmentary, he regards as absolutely reliable, a transcript by a scribe with known high accuracy but yet handwriting difficult to read might be rated 0.8, while a transcript by a scribe with neat handwriting but known careless might be rated 0.5. This rating depends on our background knowledge of the reliability of the sources. But once the sources have been ranked, the conflicting statement vouched for by the highest ranked source is recognized as the most plausible. (For this and the previous paragraph, compare [1], forthcoming).

There is one further issue to take into account in assessing the reliability of sources. Suppose there is a consensus among a subclass of independent observers about the relative number of counterexamples occurring. These sources are all moderately reliable. However, a source of highly rated reliability vouches for another relevant variable as producing the most counterexamples. Remember, however, that if a number of sources agree, we may count them as one source. In this way, a consensus of a number of moderately reliable sources may outweigh a source of very high reliability. Indeed, if one source represents a very wide convergence of sources, say sources vouching for established fact or theory, that source may then outweigh a number of other independent sources vouching for a contrary fact or theory.

7 Conclusion

Carrying out a canonical test of a generalization when the variables recognized relevant to counterexampling the generalization have been canonically ordered gives us a way of justifying a judgment of the strength of an argument whose warrant corresponds to the generalization being tested. Given this ordering, the more relevant variables whose values fail to constitute counterexamples to the generalization, the stronger the generalization and its corresponding warrant. Likewise the stronger the warrant, the stronger the argument which instances the warrant. One point remains. How strong is strong enough? That is, how strong does a warrant have to be to justify accepting the conclusion of an argument on the basis of its premises? Cohen has addressed this point in [3, pp. 310–12; 318–23]. Addressing that question is the topic of a sequel to this paper.

References

[1] J. B. Freeman. "Inferences, Inference Rules, Generalized Conditionals, Adequate Connections." In S. Oswald (Ed.) *Argumentation and Inference: Proceedings of the 2nd European Conference on Argumentation*. London: College Publications, 2018.

[2] L. J. Cohen. *The Implications of Induction*. London: Methuen & Co Ltd., 1970.

[3] L. J. Cohen. *The Probable and the Provable*. Oxford: Clarendon Press, 1977.

[4] L. J. Cohen. *An Introduction to the Philosophy of Induction and Probability*. Oxford: Clarendon Press, 1989.

[5] Irving M. Copi and Carl Cohen. *Introduction to Logic* Twelfth Edition. Upper Saddle River, NJ: Pearson Prentice Hall, 2005.

[6] Hilary Kornblith. *Inductive Inference and Its Natural Ground*. Cambridge, MA: The MIT Press, 1993.

[7] Nicholas Rescher. *Plausible Reasoning*. Assen, The Netherlands: Van Gorcum, 1976.

[8] Nicholas Rescher. *Dialectics: A Controversy-Oriented Approach to the Theory of Knowledge*. Albany, NY: State University of New York Press, 1977.

[9] Wesley C. Salmon. *The Foundations of Scientific Inference*. Pittsburgh: University of Pittsburgh Press, 1966.

[10] Stephen E. Toulmin. *The Uses of Argument*, Cambridge University Press 1958. Updated edition 2003.

Received 7 December 2017

www.ingramcontent.com/pod-product-compliance
Lightning Source LLC
Chambersburg PA
CBHW081149090426
42736CB00017B/3247
* 9 7 8 1 8 4 8 9 0 2 7 2 5 *